Lecture Notes in Mathematics

Edited by A. Dold and B. Eckmann

1305

D. S. Lubinsky
E. B. Saff

Strong Asymptotics for Extremal Polynomials Associated with Weights on $\mathbb{R}$

Springer-Verlag
Berlin Heidelberg New York London Paris Tokyo

Authors

Doron S. Lubinsky
National Research Institute for Mathematical Sciences
C.S.I.R., P.O. Box 395, Pretoria 0001
and
Department of Mathematics, University of Witwatersrand
Johannesburg 2001, Republic of South Africa

Edward B. Saff
Institute for Constructive Mathematics, Department of Mathematics
University of South Florida, Tampa, FL 33620, USA

Mathematics Subject Classification (1980): 41A25, 42C05

ISBN 3-540-18958-0 Springer-Verlag Berlin Heidelberg New York
ISBN 0-387-18958-0 Springer-Verlag New York Berlin Heidelberg

Printed in Germany

Printing and binding: Druckhaus Beltz, Hemsbach/Bergstr.
2146/3140-543210

Abstract

Let $W(x) := e^{-Q(x)}$, where $Q(x)$ is a function of smooth polynomial growth at infinity, and for $n=0,1,2,\ldots$, and $0 < p \le \infty$, define

$$E_{np}(W) := \min_{\deg(P) < n} \| (x^n - P(x))W(x) \|_{L_p(\mathbb{R})}.$$

Further, for $a > 0$, define the contracted geometric mean

$$G[W(ax)] := \exp\left(\pi^{-1} \int_{-1}^{1} \log W(ax)\, dx / \sqrt{1-x^2} \right),$$

and for $n \ge 1$, let a_n denote the positive root of

$$a_n^n G[W(a_n x)] = \max_{a>0} a^n G[W(ax)].$$

For $1 < p < \infty$, we establish the asymptotic relation

$$\lim_{n\to\infty} E_{np}(W) / \{(a_n/2)^{n+1/p}\, G[W(a_n x)]\} = 2\sigma_p,$$

where σ_p is an explicitly given constant depending only on p. For $2 \le p < \infty$, we also obtain the asymptotics in the plane of the corresponding extremal polynomials. The class of weights to which the result applies includes

$$W(x) := \exp(-|x|^\alpha [\log(A + |x|^\alpha)]^\beta),\ \alpha > 0,\ \beta \in \mathbb{R}.$$

For a more restricted class of weights, we can also treat the cases $p = 1, \infty$. Our auxiliary results include the construction of weighted polynomial approximations, inequalities for zeros of extremal polynomials, and infinite-finite-range inequalities. For the special case $p = 2$, we use the theory of orthogonal polynomials to obtain more general results.

AMS(MOS) Classification : Primary 41A25, 42C05.
Key Words and Phrases : Extremal polynomials, Orthogonal polynomials, Extremal errors, Strong asymptotics, Weighted polynomial approximations.

TABLE OF CONTENTS

1. Introduction.

(i) A Brief Review.

It was G.Freud who in the late 1960's began to develop a theory of orthogonal polynomials and weighted approximation for weights on $(-\infty,\infty)$, often in collaboration with his erstwhile student, P.Nevai. The work of Freud and Nevai consisted of Jackson-Bernstein theorems, Markov-Bernstein inequalities, estimates for Christoffel functions and the largest zeros of orthogonal polynomials, spacing of successive zeros, and convergence of Lagrange interpolation and orthogonal expansions - see Nevai [56] for an entertaining review of these developments, and for references.

One of the questions that Freud raised is the zero distribution of orthogonal polynomials for weights on $\mathbb{R}$. For a large class of weights of the form $W(x) := e^{-Q(x)}$, where $Q(x)$ is of faster than polynomial growth at infinity, Erdös [7] showed that the contracted zero distribution is arcsine, that is, if we divide the zeros of the orthogonal polynomial of degree n for $W(x)$ by its largest zero, and if we count the number $N_n(a,b)$ of these normalized zeros in any subinterval of $(-1,1)$, then

$$\lim_{n\to\infty} N_n(a,b)/n = \pi^{-1}\int_a^b dx \,/\, \sqrt{1 - x^2}, \quad -1 \le a < b \le 1.$$

It was clear from the Hermite weight $W(x) = e^{-x^2}$, that when $Q(x)$ is of polynomial growth at infinity, one could no longer expect that the contracted zero distribution of the orthogonal polynomials for W should be arcsine. The first steps in determining what the distribution should be, were taken by Nevai and Dehesa [57], who computed its moments, and Ullman [67,68], who gave an explicit formula for the density function of the distribution. Subsequent authors have hence used the term Ullman distribution, though perhaps Nevai-Ullman distribution would be a more appropriate name.

Both Nevai and Dehesa [57] and Ullman [67,68] assumed the asymptotic behaviour of the recurrence relation coefficients in deriving their results. Without any such assumptions, Rahmanov [62] and Mhaskar and Saff [46] independently established that for weights behaving like $\exp(-|x|^{\alpha})$,$\alpha > 0$, the contracted zero distribution is <u>Ullman of order α</u>, that is

$$\lim_{n\to\infty} N_n(a,b)/n = \int_a^b v(\alpha;t)\,dt \ , \ -1 < a < b < 1,$$

where

$$v(\alpha;t) := \alpha\,\pi^{-1} \int_{|t|}^{1} y^{\alpha-1}\,(y^2 - t^2)^{-1/2}\,dy \ , \ t \in (-1,1).$$

Both [46] and [62] have exercised an important influence on subsequent developments, and several authors have generalized the conditions under which the distribution should be Ullman of order α - Goncar and Rahmanov [17], Mhaskar and Saff [48,50], Lubinsky and Saff [32]. An interesting extension to orthogonal polynomials on the rays $\{ z : |\arg(z)| = \theta \}$, has recently been undertaken by Luo and Nuttall [35], in connection with the anharmonic oscillator. W.Van Assche [70] has given weighted distribution results (each zero x is given the weight $1/(1 + x^2)$), in which the limiting distribution is independent of α.

Since the contracted zero distribution involves division by the largest zero, asymptotics for this quantity are desirable, and Freud made a conjecture in this direction [13]. Rahmanov [62] proved a substantial generalization of this conjecture, showing that if $\hat{W}(x)$ is a weight satisfying for some $\alpha > 1$,

$$\lim_{|x|\to\infty} \{ \log 1/\hat{W}(x) \} / |x|^{\alpha} = 1,$$

then the largest zero x_{1n} of the nth orthogonal polynomial for $\hat{W}^2$ satisfies

$$\lim_{n\to\infty} x_{1n}/n^{1/\alpha} = c > 0.$$

Subsequently Lubinsky and Saff [32,eqn. (3.54)] generalized Rahmanov's result as follows: Let $W := e^{-Q}$, where Q is of smooth polynomial growth at infinity -we omit the technical details, simply noting that $Q(x) = |x|^{\alpha}(\log(2 + x^2))^{\beta}$, $\alpha > 0$, $\beta \in \mathbb{R}$, is a typical example. Let $a_n = a_n(W)$ denote the positive root of the equation

$$(1.1) \qquad n = 2\,\pi^{-1}\int_0^1 a_n t\, Q'(a_n t)\, dt/\sqrt{1 - t^2}\,,$$

n large enough. The number a_n was introduced by Mhaskar and Saff in [47] and has been called the <u>Mhaskar-Saff number</u> or the <u>Mhaskar-Rahmanov-Saff number</u>. If $\hat{W}(x)$ is a weight satisfying

$$\lim_{|x|\to\infty} \{ \log 1/\hat{W}(x) \} / Q(x) = 1,$$

then the largest zero x_{1n} of the nth L_p extremal polynomial for $\hat{W}^2$ satisfies

$$\lim_{n\to\infty} x_{1n}/a_n = 1, \ 0 < p \le \infty.$$

We note that the number a_n may also be described, at least for smooth weights, as that positive number maximizing $a^n G[W(ax)]$, where

$$(1.2) \qquad G[W(ax)] := \exp(\pi^{-1}\int_{-1}^1 \log W(ax)\, dx/\sqrt{1 - x^2}), \ a > 0,$$

is a <u>contracted geometric mean of W</u>.

We remark that under very general conditions on W, say

$$\int_a^b \log W(t)\,((b - t)(t - a))^{-1/2}\, dt > -\infty\,,$$

$|a|, |b|$ large enough, one expects that the smallest and largest zeros of the nth L_p extremal polynomial for W^2 should behave asymptotically like y_n and z_n respectively, where y_n and z_n are chosen to maximize

$$\{ (z - y)/2 \}^n\, G[W((y + z)/2 + x(z - y)/2)]\,,$$

over all $-\infty < y < z < \infty$. One of the challenges facing orthogonal polynomial enthusiasts is to formulate elegant general conditions guaranteeing this type of result.

For weights such as $\exp(-x^{2k})$, k an even positive integer, and more generally when the recurrence relation coefficients have a certain asymptotic behaviour, Máté, Nevai and Totik [41,42] have shown that for fixed $k \geq 1$, the kth largest zero, x_{kn}, of the nth orthonormal polynomial satisfies for some c_k independent of n,

$$x_{kn}/a_n = 1 - c_k n^{-2/3} + o(n^{-2/3}),\ n \to \infty.$$

In this paper, for a large class of weights and for L_p extremal polynomials $(2 \leq p \leq \infty)$, we shall establish the weaker estimate

$$x_{kn}/a_n = 1 + O(((\log n)\ /\ n)^{2/3})\ ,k \text{ fixed } ,n \to \infty.$$

In the proof of the above results, an important role is played by infinite-finite range inequalities, which relate the norm of a weighted polynomial over $\mathbb{R}$ to its norm over a finite interval. Let P_n denote the set of polynomials of degree at most n. Under quite general conditions on W, Freud, Nevai and others (see [56]) showed that there exists a constant $C > 0$, such that for a given $0 < p \leq \infty$,

$$(1.3) \qquad \|PW\|_{L_p(\mathbb{R})} \leq (1 + 2^{-n})\ \|PW\|_{L_p(-Cq_n,Cq_n)}\ ,P \in P_n,$$

where q_n is the positive root of the equation

$$(1.4) \qquad n = q_n Q'(q_n),$$

uniquely defined if $xQ'(x)$ is continuous and increasing in $(0,\infty)$, with limits 0 and ∞ at 0 and ∞ respectively.

An investigation of the sharp form of inequalities such as (1.3) was begun by Mhaskar and Saff [46,48]. If $W = e^{-Q}$, where for example, Q is even and convex in $(0,\infty)$ or $Q(x) = |x|^{\alpha}$, $\alpha > 0$, they showed that

$$(1.5) \qquad \|PW\|_{L_\infty(\mathbb{R})} = \|PW\|_{L_\infty[-a_n,a_n]},\ P \in P_n,$$

where a_n is the positive root of (1.1), $n \geq 1$. Further a_n is, in an asymptotic sense, "best possible". Subsequently Mhaskar and Saff [50] obtained L_p analogues of (1.5) in a general, essentially "incomplete polynomial" setting, and Lubinsky, Mhaskar and Saff [31] gave very precise L_p analogues of (1.5). For Erdös-type weights, that is weights $W := e^{-Q}$, where Q(x) is of faster than polynomial growth at infinity, sharp inequalities of this type appear in Knopfmacher and Lubinsky [20] and Lubinsky [29].

In this paper we show that convexity of Q in (1.5) can be replaced by the weaker condition that xQ'(x) is increasing to ∞ in (0,∞), together with an integrability condition at 0, and we establish in a very precise sense, the sharpness of existing L_p infinite-finite range inequalities, $0 \leq p < \infty$.

One conjecture of Freud that has had an important influence on research in orthogonal polynomials (largely thanks to the publicity afforded it by P.Nevai) concerns the recurrence relation coefficients associated with the weight $|x|^{\rho}\exp(-|x|^{\lambda})$, $\rho > -1$, $\lambda > 0$: If $p_n(x)$ is the orthonormal polynomial of degree n for this even weight, it may be generated by the relation

$$xp_n(x) = \alpha_{n+1}p_{n+1}(x) + \alpha_n p_n(x) \ , \ n = 1,2,3,\ldots \ .$$

Freud [12] conjectured that

$$\lim_{n\to\infty} \alpha_{n+1} / n^{1/\lambda} = c_\lambda > 0.$$

Freud provided a proof for $\lambda = 4,6$ -the Hermite case $\lambda = 2$ was classical. Al. Magnus [36,37] provided a proof for λ a positive even integer and also proved an analogue for certain non-even weights. Finally Lubinsky, Mhaskar and Saff [31] gave a proof for all $\lambda > 0$ and also the analogue for more general weights.

Among the ingredients of the proof in [31], the most important was the construction of certain <u>weighted polynomial approximations</u> by Lubinsky and Saff in [32]. Another essential ingredient was the criterion for asymptotics of recurrence relation coefficients associated with weights on $\mathbb{R}$, established by Knopfmacher, Lubinsky and Nevai [21]. The latter in turn used the method of proof of Máté,

Nevai and Totik [39] of Rahmanov's theorem [60,61] on the recurrence relation coefficients associated with weights positive a.e. on the unit circle.

Several extensions and refinements of the results of [31,37] have already been obtained. For weights such as $\exp(-x^{2k} + P(x))$, k an even positive integer, and P(x) a polynomial of degree less than 2k, Máté, Nevai and Zaslavsky [43] and Bauldry, Máté and Nevai [3] have established the existence of a complete asymptotic expansion for the recurrence relation coefficients. Al. Magnus [38] has reported some evidence of finitely many terms in an asymptotic expansion for the weight $\exp(-|x|^{\lambda})$, $\lambda > 1$, and Knopfmacher and Lubinsky [20] have obtained asymptotics of the recurrence relation coefficients for Erdös type weights.

The weighted polynomial approximations mentioned above involve expressions of the form $P_n(x)W(a_n x)$, $P_n \in \mathcal{P}_n$, a_n as in (1.1), rather than the more usual $P_n(x)W(x)$ (for a detailed treatment of the latter, see Ditzian and Totik [5]). The possibility of approximation by $P_n(x)W(a_n x)$ was conjectured by Saff in [63], at least for the weights $\exp(-|x|^{\alpha})$, $\alpha > 0$, and the results of [32] largely resolved this conjecture.

One of the main ideas used in [32] (in addition to the ideas of Mhaskar and Saff [46]) was the entire function

$$G_Q(x) := 1 + \sum_{n=1}^{\infty} (x/q_n)^{2n} n^{-1/2} e^{2Q(q_n)},$$

introduced by Lubinsky in [27], where it was shown under quite general conditions on Q(x), that if

$$T(x) := 1 + xQ''(x)/Q'(x),$$

then

$$G_Q(x) = W^{-2}(x)\{\pi T(x)\}^{1/2} (1 + o(1)), \quad |x| \to \infty.$$

This asymptotic enables one to replace the non-analytic weight W by the (analytic) reciprocal of an entire function with nonnegative Maclaurin series coefficients. These functions have also been used in [19] as well as in [20], in constructing approximations by $P_n(x)W(a_nx)$ for Erdős type weights.

It seems likely that polynomial approximations of this type, in a suitably weak sense, will play an important role in the developing theory of orthogonal and extremal polynomials for weights on $\mathbb{R}$.

(ii) Aims and Results of This Monograph.

While the results of Rahmanov [62] and Mhaskar and Saff [46] correspond to nth root asymptotics of orthogonal/extremal polynomials in the plane (asymptotics for $p_n(a_nz)^{1/n}$), the asymptotics obtained by Lubinsky, Mhaskar, and Saff [31] for the recurrence relation coefficients correspond to stronger ratio asymptotics (asymptotics for $p_n(a_nz)/p_{n-1}(a_nz)$) in the plane. A still stronger asymptotic is an asymptotic for leading coefficients of orthogonal polynomials, or extremal errors, that in turn leads to asymptotics for $p_n(a_nz)$ in the plane. These strong or power asymptotics for L_p extremal errors and extremal polynomials form the main focus of this monograph.

They correspond to Szegő's theory for weights on [-1,1]: If $w(x)$ is a non-negative integrable function on [-1,1], satisfying Szegő's condition

$$\int_{-1}^{1} \log w(x)\, dx/\sqrt{1-x^2} > -\infty,$$

and if

$$p_n(w,x) = \gamma_n(w)x^n + \ldots \quad , \gamma_n(w) > 0,$$

is the orthonormal polynomial of degree n for w, satisfying

$$\int_{-1}^{1} p_n(w,x)\, p_m(w,x)\, w(x)\, dx = \delta_{mn} ,$$

Szegő [66] showed that

$$(1.6)\qquad \lim_{n\to\infty} \gamma_n(w)\, 2^{-n}\, G[w]^{1/2} = \pi^{-1/2},$$

where $G[w] := G[w(x)]$ is defined in (1.2). Further, uniformly in closed subsets of $\mathbb{C}\setminus[-1,1]$,

$$(1.7)\qquad \lim_{n\to\infty} p_n(w,z)/\{\varphi(z)^n(2\pi)^{-1/2}D^{-1}(w(\cos\phi)|\sin\phi|;\ \varphi(z)^{-1})\} = 1,$$

where $\varphi(z) := z + \sqrt{z^2-1}$ (with the usual choice of branches) is the conformal map of $\mathbb{C}\setminus[-1,1]$ onto $\{z: |z| > 1\}$, and where $D(.;z)$ is the <u>Szegö function</u>

$$(1.8)\qquad D(f(\phi);z) := \exp\left(\frac{1}{4\pi}\int_{-\pi}^{\pi} \log f(\phi)\, \frac{1+ze^{-i\phi}}{1-ze^{-i\phi}}\, d\phi\right),$$

defined for $|z| < 1$.

It seems that for a very large class of even weights $w := W^2$ on $\mathbb{R}$, for which

$$G[W(ax)] > 0, \quad a \text{ large enough},$$

one expects that (1.6) should be replaced by

$$(1.9)\qquad \lim_{n\to\infty} \gamma_n(W^2)\, 2^{-n}\, a_n^{n+1/2}\, G[W(a_n x)] = \pi^{-1/2},$$

and uniformly in closed subsets of $\mathbb{C}\setminus[-1,1]$,

(1.10)

$$\lim_{n\to\infty} p_n(W^2, a_n z)/\{a_n^{-1/2}(2\pi)^{-1/2}\varphi(z)^n D^{-1}(W^2(a_n\cos\phi)|\sin\phi|;\ \varphi(z)^{-1})\} = 1,$$

where a_n is a positive number satisfying

$$a_n^n G[W(a_n x)] = \max_{a>0}\, \{a^n\, G[W(ax)]\}.$$

This conjecture (which we formulated with hindsight) we prove for a large class of even weights $W(x) := e^{-Q(x)}$, where $Q(x)$ is of smooth polynomial growth at infinity. For noneven weights, one expects that (1.9) should be replaced by

$$(1.11) \qquad \lim_{n\to\infty} \gamma_n(W^2)2^{-n} \sup_{-\infty<a<b<\infty} ((b-a)/2)^{n+1/2} G[W((a+b)/2+x(b-a)/2)] = \pi^{-1/2},$$

with a similar asymptotic for $p_n(W^2,z)$. One can of course formulate L_p analogues, $p \neq 2$, of these conjectures. In this connection, we note that only recently have strong asymptotics been obtained for L_p extremal polynomials associated with weights on [-1,1] -see [34] for the case $1 < p \leq \infty$.

Some of our results were announced in [33] and in a note added in proof in [30]. E.A. Rahmanov announced results like those above at a conference held in Segovia, Spain in September 1986, for the weights $\exp(-|x|^\lambda)$, $\lambda > 0$, when $p = 2$. For a system of orthogonal polynomials with unbounded smooth recurrence coefficients, Van Assche and Geronimo [71] obtained strong asymptotics like that above, by cleverly modifying methods of Máté-Nevai-Totik [40]. For Erdös type weights, strong asymptotics similar to those above are in preparation [29]. Finally for weights such as $\exp(-x^4)$, $\exp(-x^6)$, and $\exp(-x^4+P(x))$, $P(x)$ a polynomial of degree less than 4, asymptotics in the plane and on $\mathbb{R}$, have previously been given by Nevai [55], and his students Sheen [64,65] and Bauldry [2], even with order estimates on the error.

One of our auxiliary results that we believe to be of independent interest are polynomial approximations of the form $P_n(x)W(a_nx)$ for more general weights than in [32], and with sharper rates of convergence. Let $\eta > 0$, $h(x)$ be analytic in a neighbourhood of [-1,1] and

$$g(x) = \begin{cases} (1 - x^2)^\eta \, h(x) & , \quad x \in (-1,1), \\ 0 & , \quad x \in \mathbb{R}\backslash(-1,1). \end{cases}$$

Let a_n be the root of (1.1) in the case $Q(x) := |x|^\lambda$, $\lambda > 0$, and let $W(x) := \exp(-|x|^\lambda)$. We show that there exist polynomials P_n of degree at most n such that

$$(1.12)\qquad \|g(x) - P_n(x)W(a_nx)\|_{L_\infty(\mathbb{R})} \le C\, n^{-2\eta/3}\,(\log n)^{1+2\eta/3},$$

provided λ is a positive even integer - see Theorem 11.2. For $\lambda > 1$, we establish a similar inequality, with $n^{-2\eta/3}$ replaced by $n^{-\theta}$, where $\theta = \theta(\eta,\lambda) > 0$. For $0 < \lambda \le 1$, we obtain a similar but weaker inequality. It seems likely that one should be able to remove the log factor altogether, but that $n^{-2\eta/3}$ is best possible - the number $n^{-2/3}$ seems to play the same role near $x = \pm 1$ in (1.12) as n^{-2} plays in approximation by ordinary polynomials on $[-1,1]$.

Among the important constituents of our proofs are the following: First, the solution of certain integral equations and the discretization of potentials, as done by Rahmanov in [62], but in a more general and detailed form. Second, the majorization of weighted polynomials as done in Mhaskar and Saff [46] and continued in Lubinsky, Mhaskar and Saff [31], with applications to infinite-finite-range inequalities. Thirdly, the entire functions constructed by Lubinsky [27] and their application to weighted approximation as in Lubinsky and Saff [32]. Finally, the explicit formula for L_p extremal polynomials and errors for certain weights on $[-1,1]$, due to Bernstein.

<u>(iii) Organization of this Monograph</u>.

The reader interested only in a statement of the main results need refer only to section 3, and possibly section 4. Our main notation is summarized in section 2, and an index of notation also appears there. The paper is fairly self contained, and so the reader should not have to refer to too many other recent papers for proofs. Sections 5,6,9,10 and 13 contain background material on integral equations, potentials and discretization thereof, and Bernstein's formula. Section 7 contains an investigation of the sharpness of infinite-finite range inequalities and section 8 contains inequalities for the largest zeros of extremal polynomials. Sections 11 and 12 contain the construction of our weighted polynomial approximations, and sections 14 to 16 contain the proofs of the asymptotics for extremal errors and extremal polynomials.

(iv) Acknowledgements.

The research of D.S. Lubinsky was partially completed during a visit to the Institute for Constructive Mathematics, University of South Florida.

The research of E.B. Saff was supported, in part, by the National Science Foundation under grant DMS-862-0098.

The authors would like to acknowledge discussions with the following people, as well as their encouragement, and provision of references : S. Bonan, J.S. Geronimo, A. Knopfmacher, A.L. Levin, Al. Magnus, A. Máté, H.N. Mhaskar, P. Nevai, J. Nuttall, H. Stahl, V. Totik, J. Ullman, and W. Van Assche. Also the authors express their gratitude to Ms. Selma Canas for her careful typing of portions of this manuscript.

2. Notation and Index of Notation.

Throughout $C, C_1, C_2, C_3, \ldots$ denote positive constants independent of n and x, and occassionally also independent of any polynomial P. The same symbol does not necessarily denote the same constant in different occurrences. We use the usual o,O notation , and also ~ in the following sense:

$$f(x) \sim g(x)$$

if there exist $C_1, C_2 > 0$ such that

$$C_1 \leq f(x)/g(x) \leq C_2 ,$$

for the relevant range of x. Similar notations will be used for sequences of real numbers and sequences of functions. The class of polynomials of degree at most n, with possibly complex coefficients, will be denoted by $\mathcal{P}_n$. The greatest integer $\leq x$ will be denoted by $\langle x \rangle$.

Let $0 < p \leq \infty$ and let W(x) be a non-negative function on $\mathbb{R}$ such that W(x) is positive on a set of positive measure and

(2.1) $$x^n W(x) \in L_p(\mathbb{R}) \ , \ n=0,1,2,\ldots, \quad \text{and} \quad \|n\|_{L_p(\mathbb{R})} > 0.$$

For n=1,2,3,... we define the <u>extremal error</u>

(2.2) $$E_{np}(W) := \inf_{P \in \mathcal{P}_{n-1}} \|(x^n - P(x))W(x)\|_{L_p(\mathbb{R})} ,$$

and we define an nth L_p <u>extremal polynomial</u>

$$T_{np}(W,x) := x^n + \ldots,$$

to be any monic polynomial of degree n satisfying

(2.3) $$\|T_{np}(W,x)W(x)\|_{L_p(\mathbb{R})} = E_{np}(W).$$

Of course T_{np} is unique for $1 < p < \infty$. We also make use of the <u>normalized extremal polynomial</u>

(2.4) $$p_{np}(W,x) := T_{np}(W,x)/E_{np}(W),$$

satisfying

(2.5) $\|p_{np}(W,x)W(x)\|_{L_p(\mathbb{R})} = 1.$

In the case p = 2, $p_{np}(W,x)$ is an <u>orthonormal polynomial</u> for the weight $W^2(x)$. We shall also use the notation

(2.6) $$p_n(W^2,x) := p_{n2}(W^2,x)$$
$$= \gamma_n(W^2)x^n + \ldots,$$

where the <u>leading coefficient</u> is

(2.7) $$\gamma_n(W^2) := 1/E_{n2}(W).$$

Of course,

(2.8) $$\int_{-\infty}^{\infty} p_n(W^2,x)\ p_m(W^2,x)\ W^2(x)\ dx = \delta_{mn}.$$

When $p \geq 1$, it is known that $T_{np}(W,x)$ has n simple zeros. We frequently use the notation

$$\infty > x_{1n}^{(p)} > x_{2n}^{(p)} > \ldots x_{nn}^{(p)} > -\infty$$

for these zeros. Further we frequently use

$$\infty > y_{1n} > y_{2n} > \ldots > y_{n+1,n} > -\infty$$

to denote the n+1 equioscillation points of $T_{n\infty}(W,x)W(x)$. Whenever convenient, we shall omit unnecessary parameters in formulas.

If $xQ'(x)$ is positive and increasing in $(0,\infty)$, with limits 0 and ∞ at 0 and ∞ respectively, for n=1,2,3,... , we define q_n to be the positive root of

(2.9) $$n = q_nQ'(q_n),$$

and the <u>Mhaskar-Rahmanov-Saff (MRS) number</u> a_n to be the positive root of

(2.10) $$n = 2\ \pi^{-1}\int_0^1 a_nt\ Q'(a_nt)\ dt\ /\sqrt{1 - t^2}\ .$$

We define the series

$$(2.11)\qquad G_Q(x) := 1 + \sum_{n=1}^{\infty} (x/q_n)^{2n}\, e^{2Q(q_n)}\, n^{-1/2} .$$

The <u>geometric mean</u> of a non-negative integrable function $w(x)$ on $[-1,1]$ is

$$(2.12)\qquad G[w] := \exp\left(\pi^{-1} \int_{-1}^{1} \log w(x)\, dx \,/\, \sqrt{1 - x^2} \right),$$

taken as 0 if the integral diverges to $-\infty$. The <u>contracted geometric mean</u> of a non-negative integrable function $W(x)$ on $\mathbb{R}$ is

$$(2.13)\qquad G[W(ax)] := \exp\left(\pi^{-1} \int_{-1}^{1} \log W(ax)\, dx \,/\, \sqrt{1 - x^2} \right),\ a > 0.$$

If $f(\phi)$ is a non-negative measurable function on $[-\pi,\pi]$ satisfying

$$\int_{-\pi}^{\pi} \log f(\phi)\, d\phi > -\infty,$$

we define the <u>Szegö function</u>

$$(2.14)\qquad D(f(\phi);z) := \exp\left(\frac{1}{4\pi} \int_{-\pi}^{\pi} \log f(\phi)\, \frac{1 + ze^{-i\phi}}{1 - ze^{-i\phi}}\, d\phi\right),$$

$|z| < 1$, satisfying [66]

$$(2.15)\qquad |D(f(\phi);e^{i\theta})|^2 = f(\theta),\quad \text{a.e. } \theta \in [-\pi,\pi],$$

the left-hand side of (2.15) being a radial or non-tangential limit. Finally,

$$(2.16)\qquad \varphi(z) := z + \sqrt{z^2 - 1},$$

denotes the conformal map of $\mathbb{C}\backslash[-1,1]$ onto $\{z: |z| > 1\}$, the branch of the square root being the usual one, and we set

$$(2.17)\qquad \sigma_p := \begin{cases} [\Gamma(1/2)\ \Gamma((p + 1)/2) \,/\, \Gamma(p/2 + 1)]^{1/p}, & 0 < p < \infty, \\ 1, & p = \infty. \end{cases}$$

Following is an index of some symbols, most of which are not defined above:

Term	Description	Place of Definition
a_n	MRS number	eqn. (2.10), p.13
$A_{n,R}$	constant	Lemma 5.3, eqn. (5.46), p.38
$B_{n,R}$	constant	Lemma 5.3, eqn. (5.39), p.37
$D(\ ;z)$	Szegö function	eqn. (2.14), p.14
$E_{np}(W)$	extremal error	eqn. (2.2), p.12
$G_Q(x)$	entire function	eqn. (2.11), p.14
$G_{Q/2}(x)$	entire function	eqn. (11.2), p.80
$G[W]$	geometric mean	eqn. (2.12), p.14
$G(\epsilon;\theta)$	subset of $\mathbb{C}$ containing $(-1,1)$	Theorem 9.1, eqn. (9.2), p.67
$L[f]$	singular operator	Lemma 5.1, eqn. (5.3), p.28
$l_{jn}(x)$	fundamental polynomial	Lemma 8.6, p.62
P_n	polynomials of deg $\leq n$	p. 4
$p_n(W^2,x)$	orthonormal polynomial	eqn. (2.6), p.13
$p_{np}(W,x)$	normalized extremal polynomial	eqn. (2.4), p.12
q_n	number	eqn. (2.9), p.13
$T(x)$	function	Lemma 12.4, eqn. (12.23), p.94
$T_n(x)$	monic polynomial	Theorem 13.1, eqn. (13.8), p.112
$T_{np}(W,x)$	monic extremal polynomial	eqn. (2.3), p.12
$U_{n,R}(x)$	majorization function	Lemma 5.3, eqn. (5.51), p.39
$V(x)$	weight function	Theorem 13.1, eqn. (13.1), p.111
$V_p(x)$	weight function	Theorem 13.1, eqn. (13.2), p.111
$\gamma_n(W^2)$	leading coefficient	eqn. (2.7), p.13
$\Gamma_n, \Gamma_{n1}, \Gamma_{n2}$	contours of integration	Definition 11.3, p.82
$\kappa_n(d\mu)$	leading coefficient	Section 16.1, p.136
λ_{jn}	Christoffel number	Lemma 8.6, p.62
$\mu_{n,R}(x)$	density function	Lemma 5.3, eqn. (5.37), p.37

$\nu_{n,R}(x)$	density function	Lemma 5.3, eqn. (5.37), p.37
$\rho_{n,R,\epsilon}$	constant	Lemma 5.3, eqn. (5.48), p.38
σ_p	constant	eqn. (2.17), p.14
$\tau_{n,R}$	constant	Lemma 5.3, eqn. (5.47), p.38
$\varphi(z)$	conformal map	eqn. (2.16), p.14
$\Phi_n(z), \phi_n(z)$	complex orthogonal polys.	eqns. (16.2), (16.3), p.136
$\chi_{n,R}$	constant	Lemma 5.3, eqn. (5.52), p.39
$\Psi(x)$	function	Lemma 12.4, eqn. (12.24), p.94

3. Statement of Main Results.

Following is our main asymptotic for $E_{np}(W)$, $1 < p < \infty$.

Theorem 3.1

Let $W(x) := e^{-Q(x)}$, where $Q(x)$ is even and continuous in $\mathbb{R}$, where $Q''(x)$ exists in $(0,\infty)$ and $Q'''(x)$ exists for x large enough, with

(3.1) $\quad Q'(x) > 0 \; , \; x \in (0,\infty),$

(3.2) $\quad C_1 \leq (xQ'(x))'/Q'(x) \leq C_2 \; , \; x \in (0,\infty),$

and

(3.3) $\quad x^2|Q'''(x)|/Q'(x) \leq C_3 \; , \; x$ large enough.

Then, for $1 < p < \infty$, with the notation of (2.12), (2.13) and (2.17),

(3.4) $\quad \lim_{n\to\infty} E_{np}(W) \,/\, \{(a_n/2)^{n+1/p}\, G[W(a_nx)]\} = 2\sigma_p,$

and for $0 < p \leq 1$,

(3.5) $\quad \limsup_{n\to\infty} E_{np}(W) \,/\, \{(a_n/2)^{n+1/p}\, G[W(a_nx)]\} \leq 2\sigma_p.$

If, further, $2 \leq p < \infty$, and for $n = 1,2,3,\ldots$,

(3.6) $\quad F_n(x) := W(a_nx)\,(1 - x^2)^{1/(2p)} \; , \; x \in [-1,1]$,

then uniformly in closed subsets of $\mathbb{C}\backslash[-1,1]$,

(3.7) $\quad \lim_{n\to\infty} T_{np}(W,a_nz)/\{(a_n\varphi(z)/2)^n G[F_n(x)]D^{-2}(F_n(\cos\phi);\varphi(z)^{-1})\} = 1,$

and

(3.8) $\quad \lim_{n\to\infty} p_{np}(W,a_nz)/\{a_n^{-1/p}\varphi(z)^n\, D^{-2}(F_n(\cos\phi);\varphi(z)^{-1})\} = (2\sigma_p)^{-1}.$

As examples of $Q(x)$ satisfying (3.1) to (3.3), we mention

(3.9) $\quad Q(x) := |x|^\alpha \; , \; \alpha > 0,$

(3.10) $\quad Q(x) := |x|^\alpha(\log(A + x^2))^\beta \; , \; \alpha > 0 \; , \beta \in \mathbb{R}$, A large enough,

(3.11) $\quad Q(x) := |x|^{\alpha\{2+\sin(\epsilon \log(\log(4+x^2)))\}}, \alpha > 0, \epsilon$ small enough.

This last $Q(x)$ varies between $|x|^{\alpha}$ and $|x|^{3\alpha}$. For the cases $p = 1,\infty$, we can prove (3.4), (3.7) and (3.8) provided we assume more:

Theorem 3.2

Let $W(x)$ be as in Theorem 3.1, with the additional assumption that either $W^{-1}(x)$ is an even entire function with non-negative Maclaurin series coefficients or

(3.12) $$\lim_{n\to\infty} a_n/n^{1/2} = 0.$$

Then (3.4) remains true for $p = 1,\infty$ and (3.7) and (3.8) remain true for $p = \infty$.

As examples, we mention $W(x) := \exp(-x^2)$ or weights as in (3.9) to (3.11) satisfying

$$\lim_{|x|\to\infty} Q(x)/x^2 = \infty,$$

for, in general a_n grows roughly like $Q^{[-1]}(n)$, where $Q^{[-1]}(x)$ denotes the inverse function of $Q(x)$. The reason for the extra restrictions in Theorem 3.2 lies in our inability to approximate functions uniformly near $x=0$ by expressions of the form $P_n(x)W(a_n x)$. We note too that if Bernstein's formula (see (13.5) below) remains true for $0 < p < 1$, then the conditions of Theorem 3.2 are apparently sufficient for (3.4) to hold true also for $0 < p < 1$.

For the weight $\exp(-|x|^{\alpha})$, $\alpha > 0$, it is possible to give a more explicit form to some of the expressions in Theorem 3.1:

Corollary 3.3

Let $\alpha > 0$ and $W_{\alpha}(x) := \exp(-|x|^{\alpha})$, and let

(3.13) $$\beta_{\alpha} := \Gamma(\alpha)^{-1/\alpha}\, 2^{1-2/\alpha}\, \Gamma(\alpha/2)^{2/\alpha}.$$

<u>Then, for</u> $1 < p < \infty$,

$$(3.14)\qquad \lim_{n\to\infty} E_{np}(W_\alpha)/\{(\beta_\alpha n^{1/\alpha}/2)^{n+1/p}\, e^{-n/\alpha}\} = 2\sigma_p .$$

<u>Further, let</u>

$$(3.15)\qquad h_\alpha(z) := \beta_\alpha^\alpha\, (2\pi)^{-1} \int_{-\pi}^{\pi} |\cos\phi|^\alpha\, \frac{1 + e^{-i\phi}z}{1 - e^{-i\phi}z}\, d\phi \quad , \qquad |z| < 1.$$

<u>Then, for</u> $2 \le p < \infty$, <u>we have uniformly in closed subsets of</u> $\mathbb{C}\backslash[-1,1]$,

$$(3.16)\qquad \lim_{n\to\infty} T_{np}(W,\beta_\alpha n^{1/\alpha} z)/\{\beta_\alpha\, (n/e)^{1/\alpha}\, \varphi(z)2^{-1}\, \exp(h_\alpha(\varphi(z)^{-1}))\}^n$$
$$= (1 - \varphi(z)^{-2})^{-1/p},$$

<u>and</u>

$$(3.17)\qquad \lim_{n\to\infty} p_{np}(W,\beta_\alpha n^{1/\alpha} z)/\{(\beta_\alpha n^{1/\alpha})^{-1/p}\, \varphi(z)^n\, \exp(n\, h_\alpha(\varphi(z)^{-1}))\}$$
$$= 2^{(1/p)-1}\, \sigma_p^{-1}\, (1 - \varphi(z)^{-2})^{-1/p}.$$

<u>If</u> $\alpha \ge 2$, <u>then</u> (3.14) <u>remains valid for</u> $p = 1,\infty$, <u>and</u> (3.16) <u>and</u> (3.17) <u>remain valid for</u> $p = \infty$.

We remark that one may evaluate $h_\alpha(z)$ in terms of hypergeometric functions - see for example [46,eqns.(4.25) and (4.31)]. We shall omit the straightforward calculations that yield Corollary 3.3 from Theorems 3.1 and 3.2, and refer the reader to [46,62].

A careful examination of the proof of (3.4) in Theorem 3.1 indicates that it is possible to obtain a rate of convergence, at least for the case when $Q(x)$ grows faster than $|x|^\alpha$, some $\alpha > 1$. The rate seems to be that the expression in the left-hand side of (3.4) is of the form $2\sigma_p(1 + O(n^{-\epsilon}))$, for some $\epsilon > 0$, but we emphasise that we have not checked the details.

In forming the polynomial approximations $P_n(x)W(a_n x)$ that lead to (3.4), there is a free polynomial of degree $o(n^{1/2})$ at the nth stage, and this enables one to treat weights other than W: For example, if $\psi^{\pm 1} \in L_r[-a,a]$ for each $r < \infty$ and $a > 0$, while $\psi(x) \to 1$ as $|x| \to \infty$, then for $1 < p < \infty$,

$$\lim_{n\to\infty} E_{np}(W\psi)/\{(a_n/2)^{n+1/p}\, G[W(a_n x)]\} = 2\sigma_p .$$

Thus the "perturbation" ψ makes no difference to the asymptotic. More generally, if $\hat{W} := W\psi$, where $\psi^{\pm 1}$ is bounded in each finite interval and does not grow too fast and is sufficiently smooth, then one can show that

$$\lim_{n\to\infty} E_{np}(\hat{W})/\{(a_n/2)^{n+1/p}\, G[\hat{W}(a_n x)]\} = 2\sigma_p ,$$

for $1 < p < \infty$. One can even allow ψ to have integrable power singularities, but for technical reasons, we can allow only weak (logarithmic) zeros in ψ, so we omit these results.

Only for the case $p = 2$, shall we discuss extensions of Theorem 3.1, since here we can obtain a reasonably simple and general extension. The reason is that orthogonal polynomials for weights on [-1,1] admit a representation in terms of a corresponding set of orthogonal polynomials on the unit circle, and for extremal errors on the unit circle, there is a lower bound that is asymptotically ($n\to\infty$) the correct one. By contrast, the lower bound for $E_{np}(W)$ that one may derive directly is asymptotically too small by a factor larger than 1. This may help to explain why even for weights on [-1,1], so little (other than nth root asymptotics) is known about the L_p extremal polynomials. Only recently have strong asymptotics been obtained [34], but under more restrictive conditions than Szegö's condition and for $1 < p \leq \infty$, not for $0 < p \leq 1$.

First a result is given in which the weight of Theorem 3.1 is multiplied by a factor something like the generalized Jacobi weights considered by Nevai [54].

<u>Theorem 3.4</u>

<u>Let</u> $W(x)$ <u>be as in Theorem 3.1, and let</u> $h(x)$ <u>be a non-negative measur-</u>

able function such that

$$\lim_{|x|\to\infty} h(x) = 1, \tag{3.18}$$

and for all a large enough

$$\int_{-1}^{1} \log h(ax)\, dx/\sqrt{1-x^2} > -\infty. \tag{3.19}$$

Further let

$$w_F(x) := \prod_{j=1}^{N} |x - z_j|^{\Delta_j}, \quad x \in \mathbb{R}, \tag{3.20}$$

where $z_j \in \mathbb{C}$, $\Delta_j \in \mathbb{R}$, and if $z_j \in \mathbb{R}$, then $\Delta_j > -1/2$, $j=1,2,...N$.

Finally let

$$\Delta := \sum_{j=1}^{N} \Delta_j, \tag{3.21}$$

and

$$\hat{W}(x) := W(x)h(x)w_F(x), \quad x \in \mathbb{R}. \tag{3.22}$$

Then

$$\lim_{n\to\infty} E_{n2}(\hat{W})/\{(a_n/2)^{n+1/2+\Delta}\, G[W(a_n x)]\} = (2\pi)^{1/2}, \tag{3.23}$$

and

$$\lim_{n\to\infty} \gamma_n(\hat{W}^2)\,(a_n/2)^{n+1/2+\Delta}\, G[W(a_n x)] = (2\pi)^{-1/2}. \tag{3.24}$$

Furthermore, uniformly in closed subsets of $\mathbb{C}\backslash[-1,1]$,

$$\lim_{n\to\infty} p_n(\hat{W}^2, a_n z)/\{a_n^{-\Delta-1/2}\, \varphi(z)^n\, D^{-2}(W(a_n\cos\phi);\varphi(z)^{-1})\} \tag{3.25}$$
$$= (2\pi)^{-1/2}\,(1+\varphi(z)^{-2})^{-\Delta}\,(1-\varphi(z)^{-2})^{-1/2}\,2^{\Delta+1/2}$$

The above result is a special case of

Theorem 3.5

Let W(x) be as in Theorem 3.1, and let $\psi(x)$ be a non-negative measur-

<u>able function satisfying the following conditions</u>:

(3.26) $$\lim_{|x|\to\infty} |\log \psi(x)| \,/\, Q(x)^{1/4} = 0;$$

<u>Given any sequence of numbers</u> $\{c_n\}_{n=1}^{\infty}$ <u>satisfying</u>

(3.27) $$\lim_{n\to\infty} c_n/a_n = 1,$$

<u>there exist polynomials</u> $S_n(x)$ <u>of degree</u> $o(n^{1/2})$, $n \to \infty$, <u>positive in</u> $(-1,1)$ <u>such that</u>

(3.28) $$\lim_{n\to\infty} \int_{-1}^{1} |\log \{S_n(x)\, \psi(c_n x)\}| \, dx/\sqrt{1 - x^2} = 0,$$

(3.29) $$\lim_{n\to\infty} \|S_n\|_{L_2[-1,1]} \exp(-n^{1/4}/\log n) = 0,$$

<u>and if</u> $\{A_n\}_1^{\infty}$ <u>is a sequence of subsets of</u> $[-1,1]$ <u>such that</u> $\operatorname{meas}(A_n) \to 0$ <u>as</u> $n \to \infty$ <u>, then</u>

(3.30) $$\lim_{n\to\infty} \{ \sup_n \|S_n(x)\, \psi(c_n x)\|_{L_2(A_n)} \} = 0.$$

<u>Finally, let</u>

(3.31) $$\hat{W}(x) := W(x)\, \psi(x) \,, \quad x \in \mathbb{R}.$$

<u>Then</u>

(3.32) $$\lim_{n\to\infty} E_{n2}(\hat{W})/\{(a_n/2)^{n+1/2}\, G[\hat{W}(a_n x)]\} = (2\pi)^{1/2},$$

<u>and</u>

(3.33) $$\lim_{n\to\infty} \gamma_n(\hat{W}^2)\, (a_n/2)^{n+1/2}\, G[\hat{W}(a_n x)] = (2\pi)^{-1/2}.$$

<u>Furthermore, uniformly in closed subsets of</u> $\mathbb{C}\backslash[-1,1]$,

(3.34) $$\lim_{n\to\infty} p_n(\hat{W}^2, a_n z)/\{a_n^{-1/2} \varphi(z)^n\, D^{-2}(\hat{W}(a_n \cos\phi)\,|\sin\phi|^{1/2}; \varphi(z)^{-1})\} = (2\pi)^{-1/2}.$$

We note that although the above results are stronger than the asymptotics for the recurrence relation coefficients in [31], nevertheless, the class of weights above does not contain that in [31]: the weights there could vanish identically on a finite interval. Still, the class of weights considered here is essentially more general, since (2.13) in [32] has been dropped. For example, $Q(x)$ of (3.11) does not satisfy (2.13) in [32].

What about asymptotics of the orthogonal polynomials on the real line? In the classical Bernstein-Szegö setting for weights $w(x)$ on $(-1,1)$, relative polynomial approximation of $w(x)(1-x^2)^{1/2}$ by $P_n \in \mathcal{P}_n$, with rate $(\log n)^{-1-\delta}$, some $\delta > 0$, yields asymptotics on all of $[-1,1]$ - see [66,Thm. 12.1.3]. While we have a better rate of approximation inside most of the relevant interval $[-a_n, a_n]$, there is a small interval of length $((\log n)/n)^{2/3} a_n$ near $\pm a_n$ in which we cannot estimate a certain quantity. Only for weights such as $\exp(-|x|^\alpha)$, $\alpha < 1$, which correspond to an indeterminate moment problem, can we obtain asymptotics on the real line, and then not near $x = 0$. It seems certain that our polynomial approximations should be good enough, but a new idea is needed. For the weights $\exp(-|x|^\alpha)$, $\alpha > 0$, E.A.Rahmanov announced asymptotics on the real line at the recent Segovia conference (Spain, November 1986).

4. Weighted Polynomials and Zeros of Extremal Polynomials.

Given a sequence of polynomials $P_n \in \mathcal{P}_n$, $n = 1,2,3,\ldots$ such that

$$\|P_n(x)W(a_nx)\|_{L_\infty(\mathbb{R})} \leq C \quad , n = 1,2,3,\ldots,$$

it follows from the majorization results of [32,46,48] that

$$\lim_{n\to\infty} P_n(x)W(a_nx) = 0,$$

uniformly in closed subsets of $|x| > 1$. What about the behaviour of $\{P_n(x)W(a_nx)\}_1^\infty$ in $(-1,1)$? In particular, can we approximate continuous functions in $(-1,1)$? This question was answered for a class of weights in [32], and here we continue in this vein:

<u>Theorem 4.1</u>

<u>Let</u> $W(x)$ <u>be as in Theorem 3.1 and let</u> $g(x)$ <u>be positive and continuous in</u> $[-1,1]$. <u>Then there exist</u> $P_n \in \mathcal{P}_n$, $n = 1,2,3,\ldots,$ <u>such that</u>

$$\lim_{n\to\infty} P_n(x)W(a_nx) = g(x), \tag{4.1}$$

<u>uniformly in compact subsets of</u> $\{x\colon 0 < |x| < 1\}$,

$$\lim_{n\to\infty} P_n(x)W(a_nx) = 0, \tag{4.2}$$

<u>uniformly in closed subsets of</u> $\{x\colon |x| > 1\}$, <u>such that</u>

$$\|P_n(x)W(a_nx)\|_{L_\infty(\mathbb{R})} \leq C,\ n=1,2,3,\ldots, \tag{4.3}$$

<u>and</u>

$$\lim_{n\to\infty} \int_{-1}^{1} \left|\log |P_n(x)W(a_nx)/g(x)|\right| dx/\sqrt{1-x^2} = 0. \tag{4.4}$$

The above special case of Theorem 12.1 strengthens some of our previous results [32]. Firstly, the logarithmic convergence (4.4) shows that we can ensure that $|P_n(x)W(a_nx)|$ is not too small near $x = \pm 1$, an essential feature in the proof of our asymptotics for $E_{np}(W)$. Secondly, our weights here are (at least asymptotically) more general - in [32], we required the condition

$$\lim_{x\to\infty} (xQ'(x))'/Q'(x) = \alpha > 0,$$

which forces

$$\lim_{|x|\to\infty} \log Q(x) / \log |x| = \alpha.$$

Following is a rate of convergence:

<u>Theorem 4.2</u>

<u>Let</u> $W(x)$ <u>be as in Theorem 3.1 with the additional assumption that</u> $\theta_n := a_n/n$, $n = 1,2,3,\ldots$ <u>satisfies</u>

(4.5) $$\lim_{n\to\infty} \theta_n = 0.$$

<u>Let</u> $h(t)$ <u>be analytic in</u> $\{t \in \mathbb{C}: |t| \le 2\}$, <u>let</u> $\eta > 0$ <u>and</u>

$$g(x) := \begin{cases} (1 - x^2)^\eta h(x) & , x \in [-1,1], \\ 0, & , x \in \mathbb{R}\backslash[-1,1]. \end{cases}$$

<u>Then there exists</u> $\epsilon > 0$ <u>and</u> $P_n \in \mathcal{P}_n$, $n = 1,2,3,\ldots$, <u>such that</u>

(4.6) $$\|g(x) - P_n(x)W(a_nx)\|_{L_\infty(\mathbb{R})} \le C\,\theta_n^\epsilon, n=1,2,3,\ldots .$$

If $W(x)$ has the property that $W^{-1}(x)$ is an even entire function with non-negative Maclaurin series coefficients, such as $W(x) := \exp(-x^{2k})$, k a positive even integer, Theorem 11.2 shows that we may replace θ_n^ϵ by $((\log n)/n)^{2\eta/3} \log n$. It seems likely that in general, the correct rate of convergence in (4.6) should be $n^{-2\eta/3}$. See sections 11 and 12 for generalizations and proofs of these results.

In the introduction, we discussed infinite-finite range inequalities / identities such as

(4.7) $$\|PW\|_{L_\infty(\mathbb{R})} = \|PW\|_{L_\infty(-a_n,a_n)}, \quad P \in \mathcal{P}_n,$$

and if $P \not\equiv 0$,

(4.8) $$|P(x)W(x)| < \|PW\|_{L_\infty(-a_n,a_n)}, |x| > a_n.$$

In [46], this type of result was shown to be sharp in the sense that the largest point y_{1n} of equioscillation of $T_{n,\infty}(W,x)W(x)$ satisfies

$$T_{n,\infty}(W,y_{1n})W(y_{1n}) = \|T_{n,\infty}(W,x)W(x)\|_{L_\infty(\mathbb{R})},$$

and

$$\lim_{n\to\infty} y_{1n} / a_n = 1.$$

Here we establish in a more precise form, the sharpness of (4.7) by showing that in an interval of length $a_n\epsilon$ near a_n, $P_n(x)W(x)$ can grow by a factor of $\exp(n\,\epsilon^{3/2})$, but no faster. In particular, in a small interval of length $\sim a_n(\log n/n)^{2/3}$ near and to the left of a_n, $P_n(x)W(x)$ can grow by a given power of n:

<u>Theorem 4.3</u>

<u>Let</u> $W(x) := e^{-Q(x)}$, <u>where</u> $Q(x)$ <u>is even and continuous in</u> $\mathbb{R}$, $Q''(x)$ <u>exists in</u> $(0,\infty)$, $Q'(x)$ <u>is positive in</u> $(0,\infty)$ <u>and</u>

(4.9) $$C_1 \le (xQ'(x))' / Q'(x) \le C_2, \quad x \in (0,\infty).$$

<u>Then</u> (4.7) <u>and</u> (4.8) <u>hold for</u> $P \not\equiv 0$. <u>These are sharp in the sense that given</u> $K > 0$, <u>there exist</u> C_3 <u>and</u> $P_n \in \mathcal{P}_n$, $n = 1,2,3,\ldots$, <u>such that</u>

(4.10) $$\|P_nW\|_{L_\infty(\mathbb{R})} = \|P_nW\|_{L_\infty[-a_n,a_n]} = 1,$$

<u>and for</u> $n = 1,2,3,\ldots$,

(4.11) $$\|P_nW\|_{L_\infty(|x| \le a_n(1 - C_3\{(\log n)/n\}^{2/3}))} \le n^{-K}.$$

More detailed results of this type appear in Theorem 7.1, and L_p analogues in Theorem 7.2. One application of the results of section 7 is to the estimation of the largest zeros of L_p extremal polynomials. Recall that if $p \ge 1$,

$$\infty > x_{1n}^{(p)} > x_{2n}^{(p)} > \ldots > x_{nn}^{(p)} > -\infty,$$

denote the n simple zeros of $T_{np}(W,x)$.

Theorem 4.4

Let $W(x)$ *be as in Theorem 4.3 ,and let* j *be a fixed positive integer. There exist* C *and* n_1 *independent of* n *and* p *such that for* $n \geq n_1$ *and* $2 \leq p \leq \infty$,

$$|x_{jn}^{(p)}/a_n - 1| \leq C(\log n/n)^{2/3}. \tag{4.12}$$

As mentioned in the introduction, one expects on the basis of results of Máté, Nevai and Totik [42], that

$$x_{jn}^{(p)}/a_n - 1 = -rn^{-2/3} + o(n^{-2/3}) , \quad n \to \infty,$$

where r depends only on j and p, at least for $p = 2$. See section 8 for further results on the largest zeros.

5. Integral Equations.

In this section, we establish certain properties of solutions of integral equations with a logarithmic kernel. In the context of weights on the real line, these equations were first considered by J.L. Ullman in [67] and by E.A. Rahmanov in [62]. Here, we need to consider more general equations, and in more detail.

Lemma 5.1

Let $f(x)$ be a continuous even function on $[-1,1]$ such that $f'(x)$ exists a.e. in $[-1,1]$ and for some $1 < p < 2$, $f'(x)/\sqrt{1 - x^2} \in L_p[-1,1]$. Then the following assertions hold:

(a) The integral equation

$$(5.1) \qquad \int_{-1}^{1} \log|x-t|g(t)dt = f(x)-2\pi^{-1}\int_0^1 f(t)/\sqrt{1 - t^2}\, dt-\log 2,$$

$x \in (-1,1)$, has a solution of the form

$$(5.2) \qquad g(t):= g(f;t):= L[f'](t)+B/(\pi \sqrt{1 - t^2}), \text{ a.e. } t\in(-1,1),$$

where

$$(5.3) \qquad L[f'](t)=2\pi^{-2}PV\int_0^1 \frac{(1 - t^2)^{1/2}}{(1 - s^2)^{1/2}} \frac{sf'(s)}{(s^2 - t^2)} ds, \text{ a.e. } t\in(-1,1),$$

where PV denotes principal value, and

$$(5.4) \qquad B := B(f) := 1 - \pi^{-1} \int_{-1}^{1} sf'(s)\, (1 - s^2)^{-1/2}\, ds.$$

Furthermore,

$$(5.5) \qquad \int_{-1}^{1} L[f'](t)\, dt = \pi^{-1} \int_{-1}^{1} sf'(s)\, (1 - s^2)^{-1/2}\, ds,$$

and

$$(5.6) \qquad \int_{-1}^{1} g(t)\, dt = 1.$$

(b) For some C_1 independent of f,

(5.7) $\|L[f'](s)(1-s^2)^{-1/2}\|_{L_p[-1,1]} \le C_1 \|f'(s)(1-s^2)^{-1/2}\|_{L_p[-1,1]}$.

(c) <u>Finally, if also the function</u> $sf'(s)$ <u>is strictly increasing in</u> $(0,1)$, <u>then</u>

(5.8) $$L[f'](t) = 2\pi^{-2}\int_0^1 \frac{(1-t^2)^{1/2}}{(1-s^2)^{1/2}} \frac{(sf'(s) - tf'(t))}{(s^2 - t^2)} ds > 0,$$

a.e. $t \in (-1,1)$.

<u>Proof</u>

(a) We remark that solutions of (5.1), perhaps with a different constant on the right-hand side, are well known and widely applied. However, we could not find a clear statement of the solution (5.2), even in standard texts [18,51], and so, for the convenience of the reader, we briefly describe its derivation from the more commonly treated singular integral equation with Cauchy kernel.

Consider the equation

(5.9) $$PV \int_{-1}^{1} \psi(s)/(t-s)\, ds = f'(t),\quad t \in (-1,1),$$

with f' continuous in $[-1,1]$. If ψ' exists and is bounded in each compact subinterval of $(-1,1)$ while ψ is integrable in $(-1,1)$, then integrating (5.9) for t from -1 to x, we obtain for $x \in (-1,1)$,

(5.10) $$\int_{-1}^{1} \log|x-s|\psi(s)ds - \int_{-1}^{1} \log(1+s)\psi(s)ds = f(x) - f(-1).$$

The integration may be justified as follows: First integrate for t from $-1+\delta$ to x, some $\delta > 0$. Split the principal value integral on the left-hand side of (5.9) into integrals over $s:|t-s| \ge \epsilon$ and $s:|t-s| < \epsilon$, for some small $\epsilon = \epsilon(t) > 0$, interchange integrals, and then use the fact that

$$PV\int_{|t-s|\le\epsilon} \frac{\psi(s)}{t-s}\, ds = \int_{|t-s|\le\epsilon} \frac{\psi(s)-\psi(t)}{t-s}\, ds = O(\epsilon),\quad \epsilon \to 0,$$

by boundedness of ψ'. Finally, let $\epsilon \to 0$ and then let $\delta \to 0+$. Thus every sufficiently smooth solution ψ of (5.9) generates a solution of

(5.10).

Now let us consider (5.9). If, for example, f''' is continuous in $[-1,1]$, it is known that for any real number C

$$(5.11)\qquad \psi(t) = \frac{(1+t)^{1/2}}{\pi^2(1-t)^{1/2}} \, PV\int_{-1}^{1} \frac{(1-s)^{1/2}}{(1+s)^{1/2}} \frac{f'(s)}{(s-t)}\, ds + \frac{C}{(1-t^2)^{1/2}},$$

is a solution of (5.9). See for example [51,p.249, eqn.(88.1) and p.251, eqn.(88.8)]. Note that in our case, the parameters in [51] are $p = q = 1$ and $c_1 = -1$, $c_2 = 1$. It is easily seen from the continuity of f''' that ψ' exists and is continuous in $(-1,1)$. Thus, when f''' is continuous in $[-1,1]$, ψ also solves the equation (5.10). If we assume, in addition that f is even, so that f' is odd, then we see from (5.11) that for $t \in (0,1)$,

$$(5.12)\qquad \psi(t) = 2\pi^{-2}(1-t^2)^{1/2}\, PV\int_{0}^{1} \frac{sf'(s)}{(1-s^2)^{1/2}(s^2-t^2)}\, ds + C\,(1-t^2)^{-1/2},$$

$$(5.13)\qquad = L[f'](t) + C\,(1-t^2)^{-1/2}.$$

Using the fact that f' is odd again and (5.12), we see that also

$$(5.14)\qquad L[f'](t) = \pi^{-2}(1-t^2)^{1/2}\, PV\int_{-1}^{1} \frac{f'(s)}{(1-s^2)^{1/2}(s-t)}\, ds.$$

Now let us set $C := B/\pi$, so that $\psi(t) = g(f;t)$ is given by (5.2). Noting that [46,p.217,Lemma 4.2],

$$(5.15)\qquad \int_{-1}^{1} \log|x-s|/(\pi\,(1-x^2)^{1/2})\, dx = -\log 2,\quad s \in (-1,1),$$

and multiplying (5.10) by $1/(\pi\,(1-x^2)^{1/2})$ and integrating from -1 to 1, we obtain

$$(5.16)\qquad (-\log 2)\int_{-1}^{1} \psi(s)\, ds - \pi^{-1}\int_{-1}^{1} f(x)\,(1-x^2)^{-1/2}\, dx$$

$$= \int_{-1}^{1} \log(1+s)\,\psi(s)\, ds - f(-1).$$

Here, by (5.2),

$$\int_{-1}^{1} \psi(s)\,ds = \int_{-1}^{1} L[f'](s)\,ds + B.$$

Next, with the form (5.14) of $L[f'](t)$, Rahmanov [62,p.168] shows that (5.5) is valid by expanding the analytic function, corresponding to the principal value integral $L[f'](s)$, about ∞. Then, taking account of (5.4), we obtain

$$\int_{-1}^{1} g(s)\,ds = \int_{-1}^{1} \psi(s)\,ds = 1. \tag{5.17}$$

Using (5.16) and (5.17) to substitute for $\int_{-1}^{1} \log(1+s)\,\psi(s)\,ds - f(-1)$ in (5.10), we obtain (5.1). Thus we have established (5.2) to (5.6) of (a) in the special case that f''' is continuous in $[-1,1]$.

In proving (a) in the general case, we first need:

(b) Note that from (5.14),

$$\pi^2\, L[f'](t) \,/\, \sqrt{1-t^2} = PV \int_{-1}^{1} \frac{f'(s)}{(1-s^2)^{1/2}(s-t)}\,ds$$

is the Hilbert transform on $\mathbb{R}$ of the function defined as $f'(t)/\sqrt{1-t^2}$ on $(-1,1)$ and 0 elsewhere. A theorem of M.Riesz asserts that the Hilbert transform is a bounded operator from $L_p(\mathbb{R})$ to $L_p(\mathbb{R})$ [16,pp.128-129], and so (5.7) is valid whenever $f'(t)/\sqrt{1-t^2} \in L_p[-1,1]$.

We now return to the proof of (a) in the general case. Let $f(x)$ satisfy the hypotheses of Lemma 5.1(a), and let $F'''(x)$ be continuous in $[-1,1]$. Since (5.1) is valid for F,

$$\left|\int_{-1}^{1} \log|x-t|\,g(f;t)\,dt - f(x) + 2\pi^{-1}\int_{0}^{1} f(t)(1-t^2)^{-1/2}dt + \log 2\right| \tag{5.18}$$

$$= \left|\int_{-1}^{1} \log|x-t|\,\{g(f;t)-g(F;t)\}\,dt + F(x) - f(x)\right.$$

$$\left. + 2\pi^{-1}\int_{0}^{1} (f(t)-F(t))\,(1-t^2)^{-1/2}dt\right|$$

$$\le \left(\int_{-1}^{1} |\log|x-t||^q dt\right)^{1/q} \{\|L[f'-F']\|_{L_p[-1,1]} + C_1 |B(f)-B(F)|\}$$

$$+ 2\|f - F\|_{L_\infty[-1,1]},$$

by Hölder's inequality with $q^{-1} + p^{-1} = 1$, and by (5.2), with

$$C_1 := \|\{\pi (1 - t^2)^{1/2}\}^{-1}\|_{L_p[-1,1]}.$$

We can choose F to be a polynomial satisfying

$$\|f - F\|_{L_\infty[-1,1]} < \epsilon$$

and

$$\|L[f'-F'](t)(1 - t^2)^{-1/2}\|_{L_p[-1,1]} < \epsilon,$$

for arbitrary $\epsilon > 0$, in view of (5.7). Then it follows from (5.18) and the definition (5.4) of $B(f), B(F)$ that the right-hand side of (5.18) may be made arbitrarily small for $x \in (-1,1)$. Thus the left-hand side of (5.18) is 0 and so (5.2) to (5.4) still define a solution of (5.1). By a similar approximation argument, we see that (5.5) and (5.6) persist.

(c) Formally differentiating (5.15), we obtain

$$(5.19) \qquad PV \int_{-1}^{1} (s - x)^{-1} \pi^{-1} (1 - x^2)^{-1/2} dx = 0, \; s \in (-1,1).$$

The differentiation may be justified using the fact that $(1 - x^2)^{-1/2}$ is continuously differentiable in (-1,1), much as we justified the passage from (5.9) to (5.10). Alternatively, see [18,p.112]. From (5.19), we deduce that

$$PV \int_{0}^{1} s (s^2 - x^2)^{-1} \pi^{-1} (1 - x^2)^{-1/2} dx = 0, \; s \in (-1,1).$$

Then (5.8) follows from (5.3). If $sf'(s)$ is increasing in (0,1), then $sf'(s) - tf'(t)$ has the same sign as $s - t$, and so $L[f'](t) > 0$, a.e. $t \in (0,1)$. Further, $L[f'](t)$ is clearly even. The existence of the integral in (5.8) as an ordinary Lebesgue integral then follows from Lebesgue's Monotone Convergence Theorem and the positivity of the integrand. □

We shall need further properties of the solutions.

Lemma 5.2

Let $f(x)$ be a continuous even function such that $f'(x)$ exists a.e. in $(-1,1)$ and for some $1 < p < 2$, $f'(x)(1-x^2)^{-1/2} \in L_p[-1,1]$. Assume further that $xf'(x)$ is positive and strictly increasing in $(0,r)$ and for some $r > 1$, $f''(x)$ is continuous in $[1/2,r]$. Let

(5.20) $\quad \tau := f'(1) + \max\{|f''(u)| : u \in [1/2,1]\},$

and

(5.21) $\quad A := 2\pi^{-2} \int_0^1 \{f'(1) - tf'(t)\}\,(1-t^2)^{-3/2}\,dt,$

and let B and $L[f'](t)$ be as in Lemma 5.1. Then

(a) For some C independent of f and x, and for $x \in [7/8,1)$

(5.22) $\quad |L[f'](x)\,(1-x^2)^{-1/2} - A| \le C\,(1-x)^{1/5}\,\tau.$

Further,

(5.23) $\quad \int_{-1}^1 L[f'](x)\,(1-x)^{-1}\,dx = f'(1).$

(b) For $x \in (-1,r)$, let

(5.24) $$U(x) := \int_{-1}^1 \log|x-t|\,g(f;t)\,dt - f(x) + 2\pi^{-1}\int_0^1 f(t)\,(1-t^2)^{-1/2}\,dt + \log 2.$$

Then, letting $\rho_\epsilon := \max\{|f''(u)|: u \in [1,1+\epsilon]\}$, we obtain as $\epsilon \to 0+$,

(5.25) $\quad U'(1+\epsilon) = -A\pi(2\epsilon)^{1/2} + B((1+\epsilon)^2-1)^{-1/2} + O(\tau\epsilon^{2/3}) + O(\rho_\epsilon \epsilon),$

and

(5.26) $\quad U(1+\epsilon) = -A\pi\sqrt{8}\,\epsilon^{3/2}/3 + B\log\varphi(1+\epsilon) + O(\tau\epsilon^{5/3}) + O(\rho_\epsilon \epsilon^2),$

where the constants are independent of f and ϵ. Further, if $B \ge 0$, then

(5.27) $\quad (x\, U'(x))' < 0, \quad x \in (1,r).$

<u>Proof</u>

(a) Let $\eta := (1-x)^{2/5}$, so that $\eta \le 1/2$ for $x \in [7/8,1]$. From (5.8) and (5.21), we obtain

$$|L[f'](x)(1-x^2)^{-1/2} - A|$$

$$= 2\pi^{-2}\left|\int_0^1 \left\{\frac{sf'(s) - xf'(x)}{(1-s^2)^{1/2}(s^2-x^2)} - \frac{sf'(s) - f'(1)}{(1-s^2)^{1/2}(s^2-1)}\right\} ds\right|$$

$$\le 2\pi^{-2}\int_0^{1-\eta} (1-s^2)^{-1/2} \left| \frac{\{sf'(s) - xf'(x)\}\{x^2-1\}}{(s^2-x^2)(s^2-1)}\right.$$

$$\left. - \frac{xf'(x) - f'(1)}{s^2-1}\right| ds + 2\pi^{-2}\int_{1-\eta}^1 \frac{4\max\{|(uf'(u))'| : u \in [1/2,1]\}}{(1-s^2)^{1/2}} ds$$

$$\le C_1 f'(1)(1-x)\eta^{-2} + C_2 \max\{f''(u) : u \in [1/2,1]\}(1-x)\eta^{-1}$$

$$+ C_3\, \tau\, \eta^{1/2},$$

where C_1, C_2 and C_3 are independent of f and x. Here we have used the monotonicity of $xf'(x)$ and also that if $0 \le s \le 1-\eta$,

$$|s - x| \ge x - (1-\eta) = \eta - \eta^{5/2} \ge C_4\, \eta,$$

where $C_4 > 0$. Taking account of the definition of η and τ, we obtain (5.22).

To prove (5.23), we first assume that $f''(x)$ is continuously differentiable in $[-1,1]$. Then from (5.8), it is not difficult to see that $L[f'](t)$ is continuously differentiable in $(-1,1)$. As in the proof of Lemma 5.1, more specifically by (5.9),(5.13) and (5.19),

$$PV\int_{-1}^1 \frac{L[f'](t)}{x-t}\, dt = f'(x),\ x \in (-1,1).$$

We may write this in the form

$$(5.28) \qquad \int_{-1}^1 \frac{L[f'](t) - L[f'](x)}{x-t} dt + L[f'](x)\log\left(\frac{1+x}{1-x}\right) = f'(x),$$

$x \in (-1,1)$. By (5.22), the second term on the left-hand side of (5.28) tends to 0 as $x \to 1-$. Since $L[f'](t)$ is differentiable in $(-1,1)$ and satisfies (5.22), and since f''' is continuous, it is not difficult to see from (5.8) that $L[f'](t)$ satisfies a Lipschitz condition of order 1/2, that is

$$|L[f'](t) - L[f'](x)| \le C\,|x - t|^{1/2}, \quad x,t \in [-1,1].$$

Hence, given $0 < \eta < 1$, we have uniformly for $x \in [-1,1]$,

$$\int_{[-1,1]\cap(x-\eta,x+\eta)} \left|\frac{L[f'](t) - L[f'](x)}{x - t}\right| dt \le \hat{C}\,\eta^{1/2}.$$

Using this last estimate in (5.28), letting $x \to 1-$, and then letting $\eta \to 0+$, we obtain (5.23) in the special case that f has a continuous third derivative in $[-1,1]$. When f satisfies only the hypotheses of the lemma, we show that (5.23) persists by an approximation argument:

Let F(x) satisfy the same hypotheses as f(x) and in addition, let $F'''(x)$ be continuous in $[-1,1]$. Then

$$\left|\int_{-1}^{1} \frac{L[f'](x)}{1 - x}dx - f'(1)\right| = \left|\int_{-1}^{1} \frac{L[f' - F'](x)}{1 - x}dx - (f-F)'(1)\right|$$

$$\le \int_{-1}^{1} \frac{|L[f' - F'](x)|}{(1 - x^2)^{1/2}}\,dx + |(f - F)'(1)|$$

$$\le 2\,C_1 \|(f' - F')(x)\,(1 - x^2)^{-1/2}\|_{L_p[-1,1]} + |(f' - F')(1)|,$$

by (5.7). Since we may choose F so that this last right-hand side is arbitrarily small, (5.23) follows.

(b) We first note [46,p.217,Lemma 4.2] that

$$(5.29) \qquad \int_{-1}^{1} \log|x - t|\,\pi^{-1}(1 - t^2)^{-1/2}\,dt = \log|\varphi(x)/2|, \; x \in \mathbb{C},$$

where, as usual, $\varphi(x) := x + (x^2 - 1)^{1/2}$. Then from (5.2) and (5.4), for $x \in (1,r)$,

$$U'(x) = \int_{-1}^{1} L[f'](t)/(x - t)\,dt - f'(x) + B\,\varphi'(x)/\varphi(x).$$

Using (5.23), we then obtain for $x \in (1,r)$,

$$(5.30) \qquad U'(x) = \int_{-1}^{1} \frac{L[f'](t)(1 - x)}{(x - t)\,(1 - t)}dt + f'(1) - f'(x) + B(x^2-1)^{-1/2}.$$

Here

$$(5.31) \qquad |f'(1) - f'(x)| \le \max\{|f''(u)|: u \in [1,x]\}\,(x - 1).$$

Now let $x := 1 + \epsilon$, $\epsilon > 0$ and $\eta := \epsilon^{1/3}$. To estimate the integral on the right-hand side of (5.30), we split it into two parts. Firstly, using (5.23),

$$(5.32)\qquad \left|\int_{-1}^{1-\eta} \frac{L[f'](t)(1-x)}{(x-t)\,(1-t)}\,dt\right| \le (\epsilon/\eta)\int_{-1}^{1-\eta}\frac{L[f'](t)}{1-t}\,dt$$

$$\le \epsilon^{2/3}\, f'(1).$$

Next, by (5.22), and then by the substitution $1 - t = \epsilon u$,

$$\int_{1-\eta}^{1} \frac{L[f'](t)(1-x)}{(x-t)\,(1-t)}\,dt$$

$$= (1-x)A\int_{1-\eta}^{1}\frac{(1-t^2)^{1/2}}{(x-t)(1-t)}dt + O(\tau(x-1)\int_{1-\eta}^{1}\frac{(1-t)^{7/10}}{(x-t)(1-t)}dt)$$

$$= -\epsilon A\int_{0}^{\eta/\epsilon}\frac{(2\epsilon u-(\epsilon u)^2)^{1/2}}{\epsilon(u+1)\,\epsilon u}\,\epsilon\,du + O(\tau\epsilon\int_{0}^{\eta/\epsilon}\frac{(\epsilon u)^{-3/10}}{\epsilon(u+1)}\,\epsilon\,du)$$

$$= -A(2\epsilon)^{1/2}\int_{0}^{\eta/\epsilon}\frac{1+O(\eta)}{(u+1)u^{1/2}}\,du + O(\tau\,\epsilon^{7/10})$$

$$= -A\,(2\epsilon)^{1/2}\,(\pi + O((\eta/\epsilon)^{-1/2}) + O(\eta)) + O(\tau\,\epsilon^{7/10}),$$

by a standard integral [6,p.213,no.856.02]. Note that all the constants in the order terms are independent of f and ϵ, as the same was true in (5.22). Now it follows from (5.21) that

$$(5.33)\qquad A \le C_5\tau,$$

where C_5 is independent of f. Hence

$$(5.34)\qquad \int_{1-\eta}^{1}\frac{L[f'](t)(1-x)}{(x-t)\,(1-t)}\,dt = -A\,(2\epsilon)^{1/2}\,\pi + O(\tau\,\epsilon^{7/10}).$$

Combining (5.30),(5.31),(5.32) and (5.34), we obtain (5.25). Then (5.26) follows on integrating (5.25), and using the fact that $U(1) = 0$ (by continuity at $x = 1$ of both sides of (5.1)). Finally, to prove (5.27), we note that

$$(xU'(x))' = \frac{d}{dx}\left\{\int_{-1}^{1} L[f'](t)\frac{x}{x-t}dt - xf'(x) + Bx(x^2-1)^{-1/2}\right\}$$

$$= -\int_{-1}^{1} L[f'](t)t(x-t)^{-2}dt - (xf'(x))' - B(x^2-1)^{-3/2}.$$

Here, since f" is continuous and xf'(x) is increasing, $(xf'(x))' \geq 0$. Further, as $L[f'](t)$ is even, the integral on the last right-hand side may be expressed in the form

$$\int_0^1 L[f'](t)\, t\, \{(x - t)^{-2} - (x + t)^{-2}\}\, dt > 0$$

for $x > 1$. □

We shall need to apply the above lemmas to $f(x) := Q(Rx)/n$, and so in the following lemma, we introduce extra notation and restate parts of the above lemmas in the new notation.

Lemma 5.3

Let $W(x) := e^{-Q(x)}$, where $Q(x)$ is even and continuous in $\mathbb{R}$, $Q'(x)$ exists and is positive in $(0,\infty)$, while $xQ'(x)$ is strictly increasing in $(0,\infty)$, with

(5.35) $$\lim_{x\to\infty} xQ'(x) = \infty,$$

and for some $1 < p < 2$,

(5.36) $$\|Q'(u)\|_{L_p[0,1]} < \infty.$$

Let $a_n = a_n(W)$ for $n=1,2,3,\ldots$.

(a) Let $n \geq 1$ and $0 < R \leq a_n(W)$. Define

(5.37) $$\mu_{n,R}(x) := \upsilon_{n,R}(x) + B_{n,R}\, \pi^{-1}(1 - x^2)^{-1/2}, \quad \text{a.e. } x \in (-1,1),$$

where

(5.38) $$\upsilon_{n,R}(x) := 2\pi^{-2} \int_0^1 \frac{(1 - x^2)^{1/2}}{(1 - s^2)^{1/2}} \frac{(RsQ'(Rs) - RxQ'(Rx))}{n(s^2 - x^2)}\, ds$$

and

(5.39) $$B_{n,R} := 1 - 2\,(n\pi)^{-1} \int_0^1 \frac{RtQ'(Rt)}{(1 - t^2)^{1/2}}\, dt.$$

Then

(5.40) $$\mu_{n,R}(x) \geq \upsilon_{n,R}(x) > 0, \text{ a.e. } x \in (-1,1),$$

and

(5.41) $$\int_{-1}^{1} \mu_{n,R}(x)\, dx = 1,$$

and

(5.42) $$\|\mu_{n,R}\|_{L_p[-1,1]} \le C_1 \|RQ'(Rt)n^{-1}(1-t^2)^{-1/2}\|_{L_p[-1,1]} + C_2 B_{n,R},$$

where C_1 *and* C_2 *are independent of* n, Q *and* R. *Further,*

(5.43) $$0 \le B_{n,R} \le 1,$$

and

(5.44) $$B_{n,R} = 0 \text{ iff } R = a_n,$$

and

(5.45) $$\int_{-1}^{1} \log |x - t|\, \mu_{n,R}(t)\, dt$$

$$= Q(Rx)/n - 2(n\pi)^{-1}\int_{0}^{1} Q(Rt)(1 - t^2)^{-1/2} dt - \log 2,\ x \in [-1,1].$$

(b) *Assume in addition that* $Q''(x)$ *exists and is continuous in* $(0,\infty)$. *Define*

(5.46) $$A_{n,R} := 2\,(n\pi^2)^{-1} \int_{0}^{1} \frac{RQ'(R) - RtQ'(Rt)}{(1 - t^2)^{3/2}}\, dt,$$

and

(5.47) $$\tau_{n,R} := RQ'(R)/n + \max\,\{|R^2Q''(Ru)|/n\colon u \in [1/2,1]\},$$

and given $\epsilon > 0$,

(5.48) $$\rho_{n,R,\epsilon} := \max\,\{|R^2Q''(Ru)|/n\colon u \in [1, 1 + \epsilon]\}.$$

Then for some C_3 *independent of* n, R *and* Q,

(5.49) $$|\upsilon_{n,R}(x)\,(1 - x^2)^{-1/2} - A_{n,R}| \le C_3\,(1 - x)^{1/5}\,\tau_{n,R},$$

for $x \in [7/8, 1)$ *and* $0 < R \le a_n$. *Further,*

(5.50) $$\int_{-1}^{1} \upsilon_{n,R}(x)\,(1 - x)^{-1}\, dx = RQ'(R)/n.$$

(c) *Under the hypotheses of* (b), *for* $x \in \mathbb{C}$, *let*

$$(5.51) \qquad U_{n,R}(x) := \int_{-1}^{1} \log|x-t|\mu_{n,R}(t)\, dt - Q(R|x|)/n + \chi_{n,R}/n,$$

<u>where</u>

$$(5.52) \qquad \chi_{n,R} := 2\pi^{-1} \int_0^1 \frac{Q(Rt)}{(1-t^2)^{1/2}}\, dt + n \log 2.$$

<u>Then</u> as $\epsilon \to 0+$, <u>uniformly for</u> $0 < R \leq a_n$, $n \geq 1$,

$$(5.53) \qquad U'_{n,R}(1+\epsilon) = -A_{n,R}\,\pi\,(2\epsilon)^{1/2} + B_{n,R}((1+\epsilon)^2 - 1)^{-1/2} + O(\tau_{n,R}\,\epsilon^{2/3}) + O(\rho_{n,R,\epsilon}\,\epsilon),$$

<u>and</u>

$$(5.54) \qquad U_{n,R}(1+\epsilon) = -A_{n,R}\,\pi\,\sqrt{8}\,\epsilon^{3/2}/3 + B_{n,R}\log\varphi(1+\epsilon) + O(\tau_{n,R}\,\epsilon^{5/3}) + O(\rho_{n,R,\epsilon}\,\epsilon^2),$$

<u>where the constants in the order terms are independent of</u> n, ϵ, R <u>and</u> Q.
<u>Finally, for</u> $0 < R \leq a_n$ <u>and</u> $n \geq 1$,

$$(5.55) \qquad (xU'_{n,R}(x))' < 0, \quad x \in (1,\infty).$$

<u>If</u> $R = a_n$, <u>then also</u>

$$(5.56) \qquad U_{n,R}(x) < 0\ ;\ U'_{n,R}(x) < 0, \quad x > 1.$$

<u>Proof</u>

(a) All the assertions of (a), other than (5.43) and (5.44) follow from Lemma 5.1, if we set $f(x) := Q(Rx)/n$ and note that $\upsilon_{n,R} = L[f']$. Since $xQ'(x)$ is strictly increasing in $(0,\infty)$, both (5.43) and (5.44) follow from the definition of a_n.

(b),(c) These follow from Lemma 5.2, except for (5.56). When $R = a_n$, $B_{n,R} = 0$, so $U_{n,R}(1) = U'_{n,R}(1) = 0$, by (5.53),(5.54). Then (5.55) yields (5.56). □

We remark that one alternative form of (5.46) is

$$(5.57) \qquad A_{n,R} = 2\,(n\pi^2)^{-1} \int_0^1 \frac{t(RtQ'(Rt))'}{(1-t^2)^{1/2}}\, dt\ .$$

This follows from (5.46) with the aid of an integration by parts.

6. Polynomial Approximation of Potentials.

An important step in establishing nth root asymptotics for extremal polynomials, is the construction of polynomials that in a certain relative sense, approximate potentials. There is a standard method for constructing such polynomials, and in the context of weights on $\mathbb{R}$, Rahmanov [62] and Mhaskar and Saff [46] estimated the approximation power for this method. The following form is sufficient for our purposes.

Lemma 6.1

Let $\mu(t)$ *be a non-negative function on* $[-1,1]$ *such that*

$$(6.1) \qquad \int_{-1}^{1} \mu(t)\, dt = 1.$$

Let

$$(6.2) \qquad \Lambda(z) := \int_{-1}^{1} \log|z - t|\, \mu(t)\, dt, \quad z \in \mathbb{C}.$$

Suppose that there exist n_0, K *and* $\varepsilon > 0$ *such that for* $n \geq n_0$,

$$(6.3) \qquad \int_{|x-t| \leq n^{-K}} |\log|x - t||\, \mu(t)\, dt \leq \varepsilon (\log n)/(n + 1),$$

$x \in [-1,1]$. *Then if* $-1 = y_{0n} < y_{1n} < \ldots < y_{nn} < y_{n+1,n} = 1$ *are chosen so that*

$$(6.4) \qquad \int_{y_{jn}}^{y_{j+1,n}} \mu(t)\, dt = 1/(n + 1), \; j = 0,1,2,\ldots,n,$$

and if

$$(6.5) \qquad P_n(x) := \prod_{j=1}^{n} (x - y_{jn}),$$

we have for $n \geq n_0$,

$$(6.6) \qquad |P_n(x)| \leq e^{(n+1)\Lambda(x)}\, n^{K+\varepsilon}, \quad x \in \mathbb{R},$$

$$(6.7)\qquad |P_n(x)| \geq (1/4)e^{(n+1)\Lambda(x)}\ |x - y_{cn}|\ ,\ x \in [-1,1],$$

<u>where</u> $y_{cn} = y_{cn}(x)$ <u>denotes the closest zero of</u> $P_n(x)$ <u>to</u> x. <u>Furthermore</u>,

$$(6.8)\qquad |P_n(x)| \geq \frac{1}{|x|+1}\ e^{(n+1)\Lambda(x)}\ ,\ x \notin [y_{1n},y_{nn}].$$

<u>Proof</u>

First fix $x \in [y_{1n},y_{nn}]$ and choose l such that

$$y_{ln} \leq x < y_{l+1,n}.$$

Since $\log|x-t|$ is decreasing in t, for $t \leq y_{ln}$, we have for $j \leq l-1$,

$$\log|x-t| \leq \log|x-y_{jn}|\ ,\ t \in (y_{jn},y_{j+1,n}),$$

and

$$\log|x-t| \geq \log|x-y_{jn}|\ ,\ t \in (y_{j-1,n},y_{jn}).$$

Then taking account of (6.4), we obtain for $j \leq l-1$,

$$\int_{y_{jn}}^{y_{j+1,n}} \log|x-t|\cdot\mu(t)\ dt \leq \log|x-y_{jn}|\ /(n+1)$$

$$\leq \int_{y_{j-1,n}}^{y_{jn}} \log|x-t|\ \mu(t)\ dt,$$

and so

$$(6.9)\qquad \int_{y_{1n}}^{y_{ln}} \log|x-t|\ \mu(t)\ dt \leq (n+1)^{-1} \sum_{j=1}^{l-1} \log|x-y_{jn}|$$

$$\leq \int_{-1}^{y_{l-1,n}} \log|x-t|\ \mu(t)\ dt.$$

Similarly, since $\log|x-t|$ is increasing in t for $t \geq y_{l+1,n}$, we obtain

$$(6.10)\qquad \int_{y_{l+1,n}}^{y_{nn}} \log|x-t|\ \mu(t)\ dt \leq (n+1)^{-1} \sum_{j=l+2}^{n} \log|x-y_{jn}|$$

$$\leq \int_{y_{l+2,n}}^{1} \log|x-t|\ \mu(t)\ dt.$$

Adding (6.9) and (6.10), we obtain for $x \in [y_{1n},y_{nn}]$,

$$(6.11)\qquad \Lambda(x) - \left[\int_{-1}^{y_{1n}} + \int_{y_{ln}}^{y_{l+1,n}} + \int_{y_{nn}}^{1} \right] \log\ |x - t|\ \mu(t)\ dt$$

$$\leq (n+1)^{-1} \sum_{\substack{j=1 \\ j\neq l,\, l+1}}^{n} \log |x - y_{jn}| \leq \Lambda(x) - \int_{y_{l-1,n}}^{y_{l+2,n}} \log|x-t|\, \mu(t)dt.$$

Now if, say, y_{ln} is closer to x than $y_{l+1,n}$, that is, $y_{cn} = y_{ln}$, then by (6.4),

$$\int_{y_{ln}}^{y_{l+1,n}} \log|x-t|\, \mu(t)\, dt \leq \log |x - y_{l+1,n}| / (n+1).$$

Also $\log|x - t| \leq \log 2$, for $t \in [-1,1]$, so from (6.11),

(6.12) $\quad \Lambda(x)-(\log 4)/(n+1) \leq (n+1)^{-1}\{\log|P_n(x)| - \log|x-y_{cn}|\}$.

This yields (6.7) for $x \in [y_{1n},y_{nn}]$. Next,

$$(n+1)^{-1}\log |x - y_{ln}| \leq \int_{y_{l-1,n}}^{y_{ln}} \log |x - t|\, \mu(t)\, dt,$$

and

$$(n+1)^{-1}\log |x - y_{l+1,n}| \leq \int_{y_{l+1,n}}^{y_{l+2,n}} \log |x - t|\, \mu(t)\, dt,$$

and together with (6.11), we obtain

(6.13) $$(n+1)^{-1} \log|P_n(x)| \leq \Lambda(x) - \int_{y_{ln}}^{y_{l+1,n}} \log|x - t|\, \mu(t)\, dt.$$

We split the integral on the right-hand side of (6.13) into integrals over $t\colon |x - t| \leq n^{-K}$ and $t\colon |x - t| \geq n^{-K}$. In view of (6.3), the former integral is bounded in absolute value by $\epsilon(\log n)/(n+1)$. The latter integral is clearly bounded above by

$$K (\log n) \int_{y_{ln}}^{y_{l+1,n}} \mu(t)\, dt = K (\log n)/(n+1).$$

Thus

(6.14) $\quad (n+1)^{-1} \log|P_n(x)| \leq \Lambda(x) + (K + \epsilon)(\log n)/(n+1)$.

This yields (6.6), at least when $x \in [y_{1n},y_{nn}]$. Note that for $x \in [-1,y_{1n}] \cup [y_{nn},1]$, (6.8) yields (6.7) since $|x - y_{cn}| \leq 2$.

We proceed to prove (6.8) by estimating $|P_n(x)|$ when $x \notin [y_{1n},y_{nn}]$. Suppose, say, $x > y_{nn}$. Then, as before, we obtain

$$\int_{y_{1n}}^{1} \log|x-t|\, \mu(t)\, dt \le (n+1)^{-1} \sum_{j=1}^{n} \log|x-y_{jn}|$$

$$\le \int_{-1}^{y_{nn}} \log|x-t|\, \mu(t)\, dt,$$

and so

$$(6.15) \qquad \Lambda(x) - (\log(x+1))/(n+1) \le (n+1)^{-1} \log |P_n(x)|$$

$$\le \Lambda(x) - \int_{y_{nn}}^{1} \log|x-t|\, \mu(t)\, dt.$$

Here, if $x \in [y_{nn}, 1]$, we obtain as before,

$$\left| \int_{y_{nn}}^{1} \log|x-t|\, \mu(t)\, dt \right| \le (K + \varepsilon)(\log n)/(n+1),$$

while if $x \in [1,\infty)$,

$$-\int_{y_{nn}}^{1} \log|x-t|\, \mu(t)\, dt \le -\int_{y_{nn}}^{1} \log|1-t|\, \mu(t)\, dt$$

$$\le (K + \varepsilon)(\log n)/(n+1).$$

Then (6.15) yields (6.6) and (6.8) for $x > y_{nn}$. The proof for $x < y_{1n}$ is similar. □

In applying Lemma 6.1, we shall need some technical estimates:

<u>Lemma 6.2</u>

<u>Let</u> $W(x) := e^{-Q(x)}$, <u>where</u> $Q(x)$ <u>is even, continuous in</u> $\mathbb{R}$, $Q''(x)$ <u>exists and is continuous in</u> $(0,\infty)$, $Q'(x)$ <u>is positive in</u> $(0,\infty)$, <u>and for some</u> $C_1, C_2 > 0$,

$$(6.16) \qquad C_1 \le (xQ'(x))' \,/\, Q'(x) \le C_2\ , \quad x \in (0,\infty).$$

<u>Assume the notation of Lemma 5.3.</u>

(a) <u>Then</u>

$$(6.17) \qquad Q'(1)x^{C_2-1} \le Q'(x) \le Q'(1)x^{C_1-1}\ , \quad x \in (0,1],$$

$$(6.18) \qquad Q'(1)x^{C_1-1} \le Q'(x) \le Q'(1)x^{C_2-1}\ , \quad x \in [1,\infty),$$

<u>and for</u> $t > 1$,

(6.19) $t^{C_1-1} \leq Q'(tx)/Q'(x) \leq t^{C_2-1}$, $x \in (0,\infty)$.

Further, for x large enough,

(6.20) $C_1/2 \leq (xQ'(x))/Q(x) \leq 2\,C_2$.

(b) There exist constants $C_3, C_4, \ldots, C_{10}$ independent of n and R such that

(6.21) $n \leq a_n Q'(a_n) \leq C_3 n$,

and uniformly for $a_n/2 \leq R < a_n$,

(6.22) $B_{n,R} \sim (1 - R/a_n)$.

Furthermore, uniformly for $0 < R \leq a_n$, $n \geq 1$,

(6.23) $A_{n,R} \sim RQ'(R)/n$,

and uniformly for $a_n/2 \leq R \leq a_n$, $n \geq 1$,

(6.24) $A_{n,R} \sim 1$.

Uniformly for $0 < R \leq a_n$, $n \geq 1$ and $\varepsilon \in [0,1]$,

(6.25) $\rho_{n,R,\varepsilon} \leq C_4\, RQ'(R)/n$,

and

(6.26) $\tau_{n,R} \leq C_4\, RQ'(R)/n$.

(c) There exists $1 < p < 2$ and there exists C_6 such that uniformly for $0 < R \leq a_n$, $n \geq 1$,

(6.27) $\|\mu_{n,R}\|_{L_p[-1,1]} \leq C_6$,

and

(6.28) $\|Q'\|_{L_p[0,1]} < \infty$.

(d) There exist C_7, C_8, C_9, C_{10} and ε_0 such that uniformly for $0 < \varepsilon < \varepsilon_0$, $a_n/2 \leq R \leq a_n$, $n \geq 1$,

$$(6.29)\qquad -C_7\epsilon^{3/2} + C_8\epsilon^{1/2}(1 - R/a_n) \le U_{n,R}(1+\epsilon)$$
$$\le -C_9\epsilon^{3/2} + C_{10}\epsilon^{1/2}(1-R/a_n).$$

Proof.

(a) If $r,s > 0$, then

$$rQ'(r)\ /\ (sQ'(s)) = \exp\left(\int_s^r (uQ'(u))'/(uQ'(u))\,du\right).$$

By choosing r,s in a suitable way, and by using (6.16), we obtain (6.17) to (6.19). To obtain (6.20), integrate (6.16).

(b) By monotonicity of $uQ'(u)$, and the definition of a_n,

$$(a_n/2)Q'(a_n/2)\ \{2\pi^{-1}\int_0^{1/2} dt/\sqrt{1-t^2}\} \le n$$
$$\le a_nQ'(a_n)\ \{2\pi^{-1}\int_0^1 dt/\sqrt{1-t^2}\}.$$

Thus, for some C,

$$C\,(a_n/2)Q'(a_n/2) \le n \le a_nQ'(a_n).$$

In view of (6.19) with $t = 2$, $x = a_n/2$, (6.21) follows.

Next from (5.39), we see that for $R > 0$, $n \ge 1$,

$$\frac{d}{dR}B_{n,R} = -\frac{2}{n\pi}\int_0^1 t\,(uQ'(u))'\big|_{u=Rt}\,dt/\sqrt{1-t^2}$$

Now, by (6.16), we see that uniformly for $R > 0$,

$$C_1\int_0^1 tQ'(Rt)\,dt/\sqrt{1-t^2} \le \int_0^1 t\,(uQ'(u))'\big|_{u=Rt}\,dt/\sqrt{1-t^2}$$
$$\le C_2\int_0^1 tQ'(Rt)\,dt/\sqrt{1-t^2}.$$

Next, using the monotonicity of $uQ'(u)$ and (6.19), we see that uniformly for $R > 0$,

$$\int_0^1 RtQ'(Rt)\,dt/\sqrt{1-t^2} \sim RQ'(R).$$

Thus, uniformly for $R > 0$, $n \ge 1$,

$$R\frac{d}{dR}B_{n,R} \sim -RQ'(R)/n.$$

Using (6.19) and (6.21), we deduce that

$$(6.30)\qquad RQ'(R) \sim n\ ,\ \text{uniformly for } a_n/2 \le R \le a_n,\ n \ge 1,$$

and so

(6.31) $\frac{d}{dR} B_{n,R} \sim -1/R$, uniformly for $a_n/2 \le R \le a_n$, $n \ge 1$.

Integrating and using $B_{n,a_n} = 0$, we obtain

$$B_{n,R} = - \int_R^{a_n} \frac{d}{ds} B_{n,s}\, ds \sim \log (a_n/R) \sim (1 - R/a_n),$$

as $\log (x/y) \sim (1 - y/x)$ for $1/2 \le y/x \le 1$. Thus (6.22) is valid. Next, (6.23) is an easy consequence of (5.57) and the assumption (6.16). Further, (6.30) then yields (6.24). Next, the nonnegativity of Q' and (6.16) yield

$$|x\, Q''(x)| \le (C_2 + 1)\, Q'(x) \ , \ x \in (0,\infty).$$

Using (5.47),(5.48) and the property (6.19) of $Q'(x)$ as well as the monotonicity of $xQ'(x)$, we deduce (6.25) and (6.26).

(c) Firstly, the existence of $p > 1$ satisfying (6.28) follows from (6.17). Next, from (5.42),

$$\|\mu_{n,R}\|_{L_p[-1,1]} \le C\{\|RQ'(Rt)\, /(n\sqrt{1 - t^2})\|_{L_p[-1,1]} + B_{n,R}\}.$$

Here by (5.43), $B_{n,R} \le 1$. Further it is a consequence of the first inequality in (6.19) (replace t by $1/t$ and set $x = Rt$) that

$$Q'(Rt) \le Q'(R)\, t^{C_1-1} \ , \ t \in [0,1] \ , \ R \in (0,\infty).$$

Then

$$\|RQ'(Rt)/(n\sqrt{1 - t^2})\|_{L_p[-1,1]} \le 2RQ'(R)n^{-1}\|t^{C_1-1}/\sqrt{1 - t^2}\|_{L_p[0,1]} \le C_{11},$$

uniformly for $0 < R \le a_n$ and $n \ge 1$, provided that $p \in (1,2)$ is chosen so that $(C_1-1)p > -1$. Thus (6.27) is valid.

(d) Note first that for small enough ϵ, $\log \varphi(1 + \epsilon) \sim \epsilon^{1/2}$. Combining (5.54) with (6.22), (6.24), (6.25) and (6.26), we then obtain (6.29). □

We can now find suitable polynomial approximations for the potentials associated with $\mu_{n,R}$, giving the main lemma of this section:

Lemma 6.3

Let $W(x) := e^{-Q(x)}$, where $Q(x)$ is even, continuous in $\mathbb{R}$, $Q''(x)$ exists in $(0,\infty)$, $Q'(x)$ is positive in $(0,\infty)$ and for some $C_1, C_2 > 0$, (6.16) holds. Then there exist C_3, C_4, C_5, C_6 independent of n, x and R, such that for $n \geq 2$, and $0 < R \leq a_n$, there exists a polynomial $\hat{P}_n(x) \in \mathcal{P}_n$ (depending on n, R and Q) with n real simple zeros in $(-1,1)$, and such that

$$(6.32) \qquad C_3 \, |x - y_{cn}| \leq |\hat{P}_n(x) W(Rx)| \leq C_4 \, n^{C_5}, \quad x \in [-1,1],$$

where $y_{cn} = y_{cn}(n,R,x)$ denotes the closest zero of $\hat{P}_n(x)$ to x. Further,

$$(6.33) \qquad C_6 \leq |\hat{P}_n(x) W(Rx)| e^{-nU_{n,R}(x)} \leq C_4 \, n^{C_5}, \quad 1 \leq |x| \leq 2.$$

Finally, the leading coefficient of $\hat{P}_n(x)$ is $\exp(\chi_{n,R})$, where $\chi_{n,R}$ is given by (5.52).

Proof

First note that if $K > 0$ and $1 < p < 2$ is such that (6.27) and (6.28) hold, and if $q > 1$ satisfies $p^{-1} + q^{-1} = 1$, then for $x \in [-1,1]$,

$$\int_{|x-t| \leq n^{-K}} |\log |x - t|| \, \mu_{n,R}(t) \, dt$$

$$\leq \left[\int_{|x-t| \leq n^{-K}} |\log |x - t||^q \, dt \right]^{1/q} \left[\int_{-1}^{1} |\mu_{n,R}(t)|^p \, dt \right]^{1/p}$$

$$\leq C_6 \left(2 \int_0^{n^{-K}} |\log u|^q \, du \right)^{1/q},$$

by (6.27). By a suitable choice of K, we can ensure that this last right-hand side is bounded above by $(\log n)/n$ for $n \geq 2$. Thus (6.3) in Lemma 6.1 is satisfied, with $n_0 = 2$, $\epsilon = 1$, and

$$\Lambda(x) := \int_{-1}^{1} \log |x - t| \, \mu_{n,R}(t) \, dt,$$

and so Lemma 6.1 yields monic polynomials $P_n(x)$ of degree n with n simple zeros in $(-1,1)$, satisfying (6.6) to (6.8). Note that an application of Hölder's inequality and (6.27) show that for some C independent of n,R,

(6.34) $|\Lambda(x)| \le C \ , \ x \in [-2,2], \ n \ge 1, \ 0 < R \le a_n .$

Next, by (5.45),

(6.35) $-n\ \Lambda(x) = -\ Q(Rx) + \chi_{n,R}\ ,\ x \in [-1,1],$

where $\chi_{n,R}$ is given by (5.52). Defining

$$\hat{P}_n(x) := e^{\chi_{n,R}}\ P_n(x),$$

we see from (6.35) that

$$\hat{P}_n(x)W(Rx) = P_n(x)e^{-n\Lambda(x)}\ ,\ x \in [-1,1],$$

and from (5.51), we deduce that

$$\hat{P}_n(x)W(Rx)e^{-nU_{n,R}(x)} = P_n(x)e^{-n\Lambda(x)}\ ,\ x \in \mathbb{R}.$$

The upper bounds in (6.32) and (6.33) then follow from (6.6) and (6.34). The lower bounds in (6.32) and (6.33) follow from (6.7),(6.8) and (6.34). □

7. Infinite-Finite Range Inequalities and Their Sharpness.

An important step in the analysis of extremal polynomials on $\mathbb{R}$ is the estimation of the norm of a weighted polynomial $\|PW\|_{L_p(\mathbb{R})}$ in terms of the norm $\|PW\|_{L_p(-c_n,c_n)}$ over a finite interval $(-c_n,c_n)$. Freud, Nevai and others (see [56]) obtained estimates that sufficed for weighted Jackson-Bernstein type theorems on $\mathbb{R}$. Subsequently, Mhaskar and Saff [46] established sharper inequalities that led to nth root asymptotics for L_p extremal polynomials, and for $p = \infty$, a particularly elegant form appears in [48]. In resolving Freud's conjecture, Lubinsky, Mhaskar, and Saff [31] further sharpened these inequalities. For Erdös type weights, analogous inequalities appear in [20,29].

Here, for Freud type weights, we continue these investigations, generalizing and sharpening earlier results and establishing their sharpness in a precise sense. Recall that $\mathcal{P}_n$ denotes the class of real polynomials of degree at most n.

Theorem 7.1

Let $W(x) := e^{-Q(x)}$, where $Q(x)$ is even, continuous in $\mathbb{R}$, $Q''(x)$ exists in $(0,\infty)$, $Q'(x)$ is positive in $(0,\infty)$, while $xQ'(x)$ is strictly increasing in $(0,\infty)$, with

(7.1) $$\lim_{x\to\infty} xQ'(x) = \infty,$$

and for some $p > 1$,

(7.2) $$\|Q'\|_{L_p[0,1]} < \infty.$$

Let $a_n = a_n(W)$ denote the positive root of (2.10) for $n \geq 1$.

(i) Then for $n \geq 1$ and $P \in \mathcal{P}_n$,

(7.3) $$\|PW\|_{L_\infty(\mathbb{R})} = \|PW\|_{L_\infty[-a_n,a_n]},$$

and if $P \not\equiv 0$,

(7.4) $\quad |PW|(x) < \|PW\|_{L_\infty(\mathbb{R})}\,, \quad |x| > a_n.$

(ii) For $n \geq 1$, $0 < R \leq a_n$ and $P \in \mathcal{P}_n$, and $x \in \mathbb{C}\backslash[-1,1]$,

(7.5) $\quad |P(x)W(R|x|)| \leq \|P(t)W(Rt)\|_{L_\infty[-1,1]} \exp(nU_{n,R}(x)),$

where $U_{n,R}(x)$ is defined by (5.51).

(iii) Assume, in addition, there exist C_1 and C_2 such that

(7.6) $\quad C_1 \leq (xQ'(x))' / Q'(x) \leq C_2, \qquad x \in (0,\infty).$

Then given $K > 0$, there exist $C,L > 0$ such that if

(7.7) $\quad \chi_n := 1 - C((\log n)/n)^{2/3}, \qquad n \geq 2,$

then for n large enough,

(7.8) $\quad n^L \geq \sup_{P \in \mathcal{P}_n} \{ \|PW\|_{L_\infty(\mathbb{R})} / \|PW\|_{L_\infty[-\chi_n a_n, \chi_n a_n]} \} \geq n^K.$

When $Q(x)$ is convex, the results of (i) and (ii) appear in [48]. Since $Q(x)$ convex implies (7.1) and (7.2), parts (i) and (ii) of the above are thus a generalisation of one of the results of [48]. The result of (iii) shows that in an interval of length $O(a_n((\log n)/n)^{2/3})$ near a_n, $(PW)(x)$, $P \in \mathcal{P}_n$, can grow like a power of n. By contrast, (6.29) can be used to show that in an interval of length $O(a_n n^{-2/3})$ near a_n, $(PW)(x)$ cannot grow faster than a constant. We note that Theorem 4.3 is an immediate consequence of Theorem 7.1 and Lemma 6.2(a),(c). Before giving the proof of Theorem 7.1 we state our L_p result.

Theorem 7.2

Let $W(x) := e^{-Q(x)}$ satisfy the hypotheses of Theorem 7.1, including (7.6). Let $0 < p < \infty$. There exist C_1, C_2, C_3 with the following properties: Whenever

(7.9) $\quad n/(\log n)^2 \geq K_n \geq C_3, \quad n=2,3,4,\ldots,$

and

$$(7.10)\qquad \rho_n := 1 + (K_n(\log n)/n)^{2/3},\ n=2,3,4,\ldots,$$

then

$$(7.11)\qquad n^{-C_1K_n} \le \sup_{P \in P_n} \{ \|PW\|_{L_p(|x|\ge\rho_n a_n)} / \|PW\|_{L_p(\mathbb{R})} \} \le n^{-C_2K_n},$$

and

$$(7.12)\qquad 1+n^{-C_1K_n} \le \sup_{P \in P_n} \{\|PW\|_{L_p(\mathbb{R})} / \|PW\|_{L_p[-\rho_n a_n, \rho_n a_n]}\} \le 1+n^{-C_2K_n}.$$

It is also possible to prove an L_p analogue of (7.8).

Proof of Theorem 7.1(ii)

This follows [46]. Let $P \in P_n$. By the definition (5.51) of $U_{n,R}(x)$, we see that $\log |P_n(z)W(R|z|)\exp(-nU_{n,R}(z))|$ is subharmonic in $\mathbb{C}\backslash[-1,1]$, has a finite limit at ∞, and so is subharmonic in $\overline{\mathbb{C}}\backslash[-1,1]$. Further, as $z \to x \in [-1,1]$,

$$\log |P_n(z)W(R|z|)\exp(-nU_{n,R}(z))| \to \log |P_n(x)W(Rx)|,$$

by (5.45), which shows that

$$U_{n,R}(x) = 0 \ ,\ x \in [-1,1].$$

The maximum principle for subharmonic functions then yields

$$\log |P_n(z)W(R|z|)\exp(-nU_{n,R}(z))| \le \log \|P_n(x)W(Rx)\|_{L_\infty[-1,1]}$$

and this yields (7.5). □

Proof of Theorem 7.1(i)

From (5.56), $U_{n,R}(x) < 0$, $x > 1$, if $R = a_n$. Then (7.3) and (7.4) follow easily from (7.5). □

Proof of Theorem 7.1(iii)

We first establish the lower bound in (7.8), and to this end, use the polynomials of Lemma 6.3. That lemma shows that given $a_n \ge R > 0$ and $n \ge 2$, there exists $\hat{P}_n \in P_n$, such that

$$(7.13)\qquad \|\hat{P}_n(x)W(Rx)\|_{L_\infty[-1,1]} \le C_4 n^{C_5},$$

and

$$(7.14)\qquad |\hat{P}_n(x)W(Rx)| \ge C_6 \exp(nU_{n,R}(x)),\ x \in [1,2],$$

where C_4, C_5 and C_6 are independent of n, x and R. Next, by Lemma 6.2(d), we have uniformly for $0 < \epsilon < \epsilon_0$, $a_n/2 \le R \le a_n$,

$$U_{n,R}(1 + \epsilon) \ge -C_7\epsilon^{3/2} + C_8\epsilon^{1/2}(1 - R/a_n).$$

Choose now, for some $C > 0$,

$$R := a_n\chi_n = a_n(1 - C((\log n)/n)^{2/3})$$

and

$$\epsilon := ((\log n)/n)^{2/3}.$$

Then

$$nU_{n,R}(1 + \epsilon) \ge -C_7 \log n + C_8 C \log n \ge (2K + 2C_5)\log n,$$

if we choose C large enough, but fixed. Defining

$$P_n^*(x) := \hat{P}_n(x/R) = \hat{P}_n(x/(a_n\chi_n)),$$

we obtain from (7.14),

$$\|P_n^* W\|_{L_\infty(\mathbb{R})} \ge C_6\, n^{2K+2C_5},$$

while from (7.13),

$$\|P_n^* W\|_{L_\infty[-\chi_n a_n, \chi_n a_n]} \le C_4 n^{C_5}.$$

Then the lower bound in (7.8) follows.

To establish the corresponding upper bound, we note that from (7.3), (7.5) and Lemma 6.2(d), for each $P \in \mathcal{P}_n$, and $R = \chi_n a_n$,

$$\|PW\|_{L_\infty(\mathbb{R})}/\|PW\|_{L_\infty[-\chi_n a_n, \chi_n a_n]}$$

$$\le \max\{\exp(nU_{n,R}(x/R)): x \in [R, a_n]\}$$

$$\le \max\{\exp[-nC_9\epsilon^{3/2} + nC_{10}\epsilon^{1/2}C((\log n)/n)^{2/3}]: \epsilon \in [0, \chi_n^{-1} - 1]\}$$

$$\le n^L,$$

some L large enough. □

In the proof of Theorem 7.2 and subsequent sections, we shall need the following crude Nikolskii inequality, in which the essential feature is the uniformity in p of the constants.

Lemma 7.3

Let $W(x) := e^{-Q(x)}$ *be even and continuous in* $\mathbb{R}$, *and let* $Q(x)$ *be strictly increasing in* $(0,\infty)$ *with*

$$\lim_{|x|\to\infty} \log Q(x)/\log|x| = \infty. \tag{7.15}$$

There exists $C > 0$ *independent of* n,p,W *and* P, *such that for* $n \geq 1$, $P \in \mathcal{P}_n$ *and* $0 < p \leq \infty$,

$$\|PW\|_{L_\infty(\mathbb{R})} \leq \{W(0)/W(2)\}\,(Cn^2)^{1/\min\{1,p\}}\,\|PW\|_{L_p(\mathbb{R})}. \tag{7.16}$$

Proof

Note first that for some $C' > 0$ independent of n and P, we have for $n \geq 1$ and $P \in \mathcal{P}_n$,

$$\|P\|_{L_\infty[-1,1]} \leq C'n^2\|P\|_{L_1[-1,1]}. \tag{7.17}$$

See for example [54]. Then an application of Hölder's inequality shows that for $1 \leq p \leq \infty$,

$$\|P\|_{L_\infty[-1,1]} \leq 2C'n^2\|P\|_{L_p[-1,1]}.$$

By noting that for $0 < p < 1$,

$$\int_{-1}^{1} |P(t)|\,dt \leq \|P\|^{1-p}_{L_\infty[-1,1]} \int_{-1}^{1} |P(t)|^p\,dt,$$

we obtain for $n \geq 1$, $P \in \mathcal{P}_n$ and $0 < p < \infty$,

$$\|P\|_{L_\infty[-1,1]} \leq (Cn^2)^{1/\min\{1,p\}}\|P\|_{L_p[-1,1]}, \tag{7.18}$$

where $C := 2C'$ is independent of n,p and P. We remark that such techniques are standard [54], and we derived (7.18) only to obtain the uniformity in p. Next let $n \geq 1$, $P \in \mathcal{P}_n$ and ξ be such that

$$\|PW\|_{L_\infty(\mathbb{R})} = |PW|(\xi).$$

Suppose first that $\xi \geq 2$. Then by the monotonicity of W, and by

(7.18),

$$\|PW\|_{L_p(\mathbb{R})} \geq \|PW\|_{L_p[\xi-2,\xi]}$$

$$\geq W(\xi)\ \|P\|_{L_p[\xi-2,\xi]}$$

$$\geq (Cn^2)^{-1/\min\{1,p\}}\ W(\xi)\ \|P\|_{L_\infty[\xi-2,\xi]}$$

$$\geq (Cn^2)^{-1/\min\{1,p\}}\ |PW|(\xi)$$

$$= (Cn^2)^{-1/\min\{1,p\}}\ \|PW\|_{L_\infty(\mathbb{R})}.$$

Next, if $0 \leq \xi \leq 2$, we obtain similarly

$$\|PW\|_{L_p(\mathbb{R})} \geq (Cn^2)^{-1/\min\{1,p\}}\ W(2)\ \|P\|_{L_\infty[0,2]}$$

$$\geq (Cn^2)^{-1/\min\{1,p\}}\ \{W(2)/W(0)\}\ |PW(\xi)|.$$

Thus (7.16) holds if $0 \leq \xi < \infty$. Similarly if $\xi < 0$. □

For sharper weighted Nikolskii inequalities, see [44,46,22,23] and most especially [59]. Next we obtain the upper bounds in (7.11) and (7.12) uniformly in p:

<u>Lemma 7.4</u>

<u>Let</u> $W := e^{-Q}$ <u>satisfy the hypotheses of Theorem 7.2. Let</u> $0 < p_0 < \infty$. <u>There exist</u> C_2 <u>and</u> C_3 <u>independent of</u> n, P <u>and</u> p <u>and</u> $\{K_n\}_{n=1}^{\infty}$ <u>such that if (7.9) and (7.10) hold, then for</u> $p_0 \leq p \leq \infty$, $P \in \mathcal{P}_n$, $n \geq 2$,

$$(7.19)\qquad \|PW\|_{L_p(|x|\geq\rho_n a_n)}/\|PW\|_{L_p(\mathbb{R})} \leq n^{-C_2K_n},$$

<u>and</u>

$$(7.20)\qquad \|PW\|_{L_p(\mathbb{R})}/\|PW\|_{L_p[-\rho_n a_n,\rho_n a_n]} \leq 1 + n^{-C_2K_n}.$$

<u>Proof</u>

We remark that when $Q(x) = |x|^{\alpha}$, $0 < \alpha < 1$, or $Q(x)$ is even and convex, this result is essentially Theorem 2.6 in [31]. By Lemma 6.2(d) with $R = a_n$,

$$(7.21)\qquad U_{n,R}(1+\epsilon) \sim -\epsilon^{3/2},\ 0 < \epsilon < \epsilon_0,$$

uniformly for $n \geq 1$. Next, by (5.53), and by (6.24),(6.25) and (6.26) with $R = a_n$,

$$U'_{n,R}(1+\epsilon) \sim -\epsilon^{1/2},\ 0 < \epsilon < \epsilon_0.$$

Further, by (5.55), $xU'_{n,R}(x)$ is decreasing for $x > 1$, so

$$U'_{n,R}(x) \leq -C_4\delta^{1/2}x^{-1},\ x \geq 1+\delta,\ 0 < \delta < \epsilon_0,$$

where C_4 is independent of δ and x. Integrating, and using (7.21), we obtain for $x \geq 1+\delta$, $0 < \delta < \epsilon_0$,

$$U_{n,R}(x) \leq -C_5\{\delta^{3/2} + \delta^{1/2}\log(x/(1+\delta))\}.$$

A straightforward calculation shows that if ρ_n is given by (7.10),

$$I_{n,p} := \{\int_{|x|\geq\rho_n a_n} \exp(npU_{n,a_n}(x/a_n))\,dx\}^{1/p}$$

$$\leq C_6\{a_n/(n\,K_n\,\log n)^{1/2}\}^{1/p}\,n^{-C_5K_n},$$

uniformly for $p_0 \leq p < \infty$ and $n \geq 2$ provided C_3 in (7.9) is large enough. Next, if $P \in \mathcal{P}_n$ and $n \geq 2$, (7.5) with $R = a_n$, shows that for $n \geq 2$,

$$\|PW\|_{L_p(|x|\geq\rho_n a_n)} \leq \|PW\|_{L_\infty[-a_n,a_n]}\,I_{n,p}$$

$$\leq \{W(0)/W(2)\}\,(Cn^2)^{1/\min\{1,p\}}\,\|PW\|_{L_p(\mathbb{R})}\,C_6a_n^{1/p}\,n^{-C_5K_n},$$

by Lemma 7.3. The constants are independent of n,p,P and $\{K_n\}$. Since (6.17), (6.18) and (6.21) in Lemma 6.2 show that

$$a_n \leq n^{C_7},\ n \geq 2,$$

we obtain for some C_8 and $p_0 \leq p < \infty$, $n \geq 2$,

$$\|PW\|_{L_p(|x|\geq\rho_n a_n)} \leq n^{C_8-C_5K_n}\,\|PW\|_{L_p(\mathbb{R})},$$

with C_8, C_5 independent of n,p,P and $\{K_n\}$. If

$$K_n \geq C_3 := 2C_8/C_5, \quad n \geq 1,$$

we then obtain (7.19) and hence also (7.20), even if $p_0 < 1$ (possibly with a smaller C_2). The inequalities for $p = \infty$ follow by letting $p \to \infty$. □

<u>Proof of the lower bounds in (7.11) and (7.12) of Theorem 7.2</u>

It suffices to prove the lower bound in (7.11); the lower bound in (7.12) then follows easily. We use the polynomials of Lemma 6.3 with $R = a_n$. Setting

$$P_n(x) := \hat{P}_n(x/a_n),$$

we obtain from (6.32) and (6.33),

$$(7.22) \qquad \|P_n W\|_{L_\infty(\mathbb{R})} = \|\hat{P}_n(x)W(a_n x)\|_{L_\infty[-1,1]} \leq C_4 n^{C_5},$$

and for $0 < \varepsilon \leq 1$,

$$|P_n W|(a_n(1+\varepsilon)) \geq C_6 \exp(\, nU_{n,a_n}(1+\varepsilon))$$

$$\geq C_6 \exp(\, -nC_7\, \varepsilon^{3/2}),$$

for $0 < \varepsilon < \varepsilon_0$, by (6.29). Hence if

$$\hat{\rho}_n := 1 + 2(K_n(\log n)/n)^{2/3}, \quad n \geq 2,$$

we obtain for $p_0 \leq p \leq \infty$,

$$(7.23) \qquad \|P_n W\|_{L_p(a_n\rho_n \leq |x| \leq a_n\hat{\rho}_n)} \geq n^{-C_9 K_n},$$

with C_9 independent of n and p. Further, by (7.20),

$$\|P_n W\|_{L_p(\mathbb{R})} \leq (1 + n^{-C_2 K_n})\, \|P_n W\|_{L_p[-a_n\rho_n, a_n\rho_n]}$$

$$\leq (1 + n^{-C_2 K_n})\, (2\,\rho_n a_n)^{1/p}\, \|P_n W\|_{L_\infty(\mathbb{R})}.$$

Then this last inequality and (7.22) and (7.23) yield the lower bound in (7.11). □

8. The Largest Zeros of Extremal Polynomials

It was G. Freud who first studied in some detail the largest zeros of orthogonal polynomials associated with weights on the real line [9,10,11,13,15] establishing several inequalities. E.A. Rahmanov [62] proved a substantial generalisation of one of Freud's conjectures in this connection, showing that if for some $\alpha > 1$,

$$\lim_{|x|\to\infty} \{\log 1/W(x)\}/|x|^{\alpha} = 1,$$

then the largest zero X_n, say, of the nth orthonormal polynomial for W satisfies

$$\lim_{n\to\infty} X_n/n^{1/\alpha} = c_{\alpha} > 0.$$

Subsequently, Lubinsky and Saff [32,Lemma 3.7] generalized Rahmanov's result to the case where $|x|^{\alpha}$ is replaced by a more general growth function Q(x). In this case, one has

$$\lim_{n\to\infty} X_n/a_n = 1,$$

where $a_n = a_n(e^{-Q})$ is the MRS number and X_n is the largest zero of $T_{np}(W,x)$, $0 < p \le \infty$.

For more special weights, such as $W(x) := \exp(-x^m)$, m a positive integer, Máté, Nevai and Totik [41,42] have established the following finer asymptotic for the largest zero of $p_n(W^2,x)$:

$$X_n/a_n = 1 - cn^{-2/3} + o(n^{-2/3}), \qquad n \to \infty,$$

and they also showed that this formula (with possibly different c) also holds for the jth largest zero of $p_n(W^2,x)$. While we cannot prove results as strong this, we can treat more general weights, and L_p extremal polynomials, $2 \le p < \infty$. Recall that the zeros of $T_{np}(W,x)$ are denoted by

$$-\infty < x_{nn}^{(p)} \le \dots \le x_{2n}^{(p)} \le x_{1n}^{(p)} < \infty.$$

Theorem 8.1

Let $W(x) := e^{-Q(x)}$, *where* $Q(x)$ *is even and continuous in* $\mathbb{R}$, $Q''(x)$ *exists and is continuous in* $(0,\infty)$, $Q'(x)$ *is positive in* $(0,\infty)$, *while for some* $C_1, C_2 > 0$,

(8.1) $$C_1 \leq (xQ'(x))'/Q'(x) \leq C_2, \quad x \in (0,\infty).$$

Let $a_n = a_n(W)$ *for* $n \geq 1$, *and let* j *be a fixed positive integer*.

(a) *There exist* C_3 *and* n_1 *independent of* n *and* p *such that for* $2 \leq p \leq \infty$ *and* $n \geq n_1$,

(8.2) $$|x_{jn}^{(p)}/a_n - 1| \leq C_3\,((\log n)/n)^{2/3}.$$

(b) *Given* $0 < r < \infty$, *there exist* C_4 *and* n_2 *independent of* n *and* p *such that for* $r \leq p \leq \infty$ *and* $n \geq n_2$,

(8.3) $$x_{1n}^{(p)}/a_n - 1 \leq C_4((\log n)/n)^{2/3}.$$

Unfortunately, our method for obtaining lower bounds does not work when $p < 2$. As a consequence of the above result, we can say something about the points of equioscillation of $T_{n\infty}(W,x)W(x)$:

Corollary 8.2

Let $W(x)$ *satisfy the hypotheses of Theorem 8.1, and let* $a_n = a_n(W)$ *for* $n \geq 1$, *and* j *be a fixed positive integer. Let*

$$-\infty < y_{n+1,n} < y_{n,n} < \dots < y_{1n} < \infty,$$

denote the points of equioscillation of $T_{n\infty}(W,x)W(x)$. *Then there exist* C *and* n_1 *such that for* $n \geq n_1$,

(8.4) $$1 - C((\log n)/n)^{2/3} \leq y_{jn}/a_n \leq 1.$$

We remark that the upper bound in (8.4) was obtained in [48] when $Q(x)$ is even and convex. Thus our upper bound is a generalization of that in [48]. Note that Theorem 4.4 is a restatement of Theorem 8.1(a).

Proof of Theorem 8.1(b).

We shall use Lemma 7.4. By that lemma, we can choose

$$K_n := K \ge C_3,$$

such that K is independent of p ($r \le p \le \infty$) and with

$$C_2K \ge 2,$$

and such that for $n \ge 2$, $P \in P_n$, $r \le p \le \infty$,

$$\|PW\|_{L_p(\mathbb{R})} \le (1 + n^{-2}) \|PW\|_{L_p[-(1+\epsilon_n)a_n,(1+\epsilon_n)a_n]}, \tag{8.5}$$

where

$$\epsilon_n := (K (\log n)/n)^{2/3}, \quad n=2,3,4,\ldots .$$

Let us suppose that for some $r \le p \le \infty$, $n \ge 2$, $T_{np}(W,x)$ has a zero $X_n = x_{n1}^{(p)}$ with

$$X_n \ge a_n(1 + 3\epsilon_n). \tag{8.6}$$

Let

$$r(x) := (x - a_n(1 + 2\epsilon_n))/(x - X_n)$$
$$= 1 + \{X_n - a_n(1 + 2\epsilon_n)\}/(x - X_n).$$

It is easy to see that

$$\|r(x)\|_{L_\infty[-(1+\epsilon_n)a_n,(1+\epsilon_n)a_n]} = |r(-a_n(1+\epsilon_n))| \le (2 + 3\epsilon_n)/(2 + 4\epsilon_n), \tag{8.7}$$

provided that n is so large that $\epsilon_n < 1$, the lower bound on n being independent of p. Note too that

$$(1 + n^{-2})(2 + 3\epsilon_n)/(2 + 4\epsilon_n) = 1 - \epsilon_n/2 + n^{-2} + O(\epsilon_n^2) < 1$$

for $n \ge n_1$, n_1 being independent of p. Now let

$$S_n(x) := T_{np}(W,x)\, r(x),$$

a monic polynomial of degree n. By (8.5), we have

$$\|S_nW\|_{L_p(\mathbb{R})} \le (1 + n^{-2}) \|T_{np}(W,x)W(x)r(x)\|_{L_p[-(1+\epsilon_n)a_n,(1+\epsilon_n)a_n]}$$

$$\le (1 + n^{-2})\, (2 + 3\epsilon_n)/(2 + 4\epsilon_n)\, \|T_{np}(W,x)W(x)\|_{L_p(\mathbb{R})}$$

$$< \|T_{np}(W,x)W(x)\|_{L_p(\mathbb{R})} .$$

if $n \geq n_1$, contradicting extremality of T_{np}. Thus

$$X_n \leq a_n(1 + 3\epsilon_n),\ n \geq n_1,\ r \leq p \leq \infty. \ \square$$

We shall need several lemmas in the proof of (8.2).

Lemma 8.3

Let $W(x) := e^{-Q(x)}$ satisfy the hypotheses of Theorem 8.1. Let $s > 0$, and

(8.8) $$g(r) := r^a(\log r)^b(\log\log r)^c(\log\log\log r)^d \ldots,$$

for r large enough, where $a, b, c, d, \ldots$ are arbitrary real numbers, of which at most finitely many are non-zero. Then there exists an entire function

(8.9) $$H(x) := \sum_{n=0}^{\infty} h_{2n}x^{2n},\ h_{2n} \geq 0,\ n=0,1,2,\ \ldots,$$

with

(8.10) $$H(x) \sim W(x)^{-s}/g(x),\ |x| \to \infty.$$

Proof

This follows from Theorem 1 in Lubinsky [27,p.299]. It is not difficult to see that (8.1) implies that $W(x)^{s/2} = \exp(-sQ(x)/2)$ is a Freud weight in the sense of Definition 1 in [27,p.299]. $\square$

Lemma 8.4

Let $d\beta(x)$ be a non-negative mass distribution on $(-\infty,-\infty)$ with all moments finite. Let $n \geq 1$, and let $x_{1n}, x_{2n}, \ldots, x_{nn}$, and $\lambda_{1n}, \lambda_{2n}, \ldots, \lambda_{nn}$ denote respectively the Gauss points and weights of order n for $d\beta$. Let $H(x)$ be an entire function of the form (8.9). Then

(8.11) $$\sum_{j=1}^{n} \lambda_{jn}H(x_{jn}) \leq \int_{-\infty}^{\infty} H(x)\ d\beta(x).$$

Proof

See for example Freud [8,p.92]. □

We remark that more general Markov-Posse-Stieltjes inequalities of this type were obtained by Knopfmacher and Lubinsky [19].

Lemma 8.5

Let $W(x)$ be as in Theorem 8.1, and let $a_n = a_n(W)$ for $n = 1,2,3,\ldots$. Then there exist n_1, C_3 and C_4 such that for $n \geq n_1$,

$$(8.12) \qquad n^{C_3} \leq a_n \leq n^{C_4}.$$

Further, uniformly for $n \geq n_1$ and $1 \leq |k| \leq n/2$,

$$(8.13) \qquad a_{n+k}/a_n - 1 \sim k/n.$$

Proof

Firstly, (8.12) is an easy consequence of (6.18) and (6.21). Next, by the definition of a_n,

$$k = 2\pi^{-1}\int_0^1 \{a_{n+k}tQ'(a_{n+k}t) - a_n tQ'(a_n t)\}\, dt \,/\, \sqrt{1 - t^2}$$

$$= 2\pi^{-1}\int_0^1 \{a_{n+k}t - a_n t\}\, (uQ'(u))'\, dt \,/\, \sqrt{1 - t^2},$$

where $u = u(t)$ lies between $a_n t$ and $a_{n+k}t$. Using (8.1), we see that this last integral lies between C_1 and C_2 multiplied by

$$2\pi^{-1}\int_0^1 \{a_{n+k}t - a_n t\}\, Q'(u)\, dt \,/\, \sqrt{1 - t^2}.$$

Since $uQ'(u)$ is increasing, we see that this last integral in turn lies between

$$2\pi^{-1}\int_0^1 \{a_{n+k}t - a_n t\}\, \{a_{n+k}tQ'(a_{n+k}t)\,/(a_n t)\}\, dt \,/\, \sqrt{1 - t^2}$$

$$= (a_{n+k}/a_n - 1)\,(n + k),$$

and

$$2\pi^{-1}\int_0^1 \{a_{n+k}t - a_n t\}\, \{a_n tQ'(a_n t)\,/(a_{n+k}t)\}\, dt \,/\, \sqrt{1 - t^2}$$

$$= (1 - a_n/a_{n+k})\, n.$$

Then (8.13) follows easily. □

Lemma 8.6

Let $W(x)$ be as in Theorem 8.1. There exist C_3, C_4, n_1 such that for $2 \leq p \leq \infty$, $n \geq n_1$ and $P \in P_{n-1}$,

$$(8.14) \qquad |P(x)| \leq C_3 n^{C_4} \max_{1 \leq j \leq n} |PW|(x_{jn}^{(p)}) \max_{1 \leq j \leq n} |p_{np}(W,x)/(x - x_{jn}^{(p)})|,$$

and for $x \in \mathbb{R}$,

$$(8.15) \qquad |PW|(x) \leq C_3 n^{C_4} \max_{1 \leq j \leq n} |PW(x_{jn}^{(p)})/(x - x_{jn}^{(p)})|.$$

Proof

Suppose first $2 \leq p < \infty$. Since $p_{np}(x) := p_{np}(W,x) = T_{np}(W,x)/E_{np}(W)$ is an extremal polynomial,

$$\int_{-\infty}^{\infty} p_{np}(x)\, \pi_{n-1}(x)\, |p_{np}(x)|^{p-2}\, W(x)^p\, dx = 0,$$

for every $\pi_{n-1} \in P_{n-1}$. Let

$$d\hat{\beta}(x) := |p_{np}(x)|^{p-2} W(x)^p\, dx.$$

Then p_{np} is an orthogonal polynomial for $d\hat{\beta}$ and, even more, an orthonormal polynomial, since

$$\int_{-\infty}^{\infty} p_{np}^2(x)\, d\hat{\beta}(x) = \int_{-\infty}^{\infty} |p_{np}|^p W^p dx = 1.$$

Let $l_{jn}(x)$, $j=1,2,\ldots,n$ denote the fundamental polynomials of Lagrange interpolation at the zeros of p_{np}. We shall use the following well known formula for $l_{jn}(x)$ [8,p.114,eqn.(6.3)] and [54,p.6]:

$$l_{jn}(x) = \lambda_{jn}(\gamma_{n-1}/\gamma_n)\, p_{n-1}(x_{jn}^{(p)})\, p_{np}(W,x)/(x - x_{jn}^{(p)}).$$

Here λ_{jn} is the jth Christoffel number of order n for $d\hat{\beta}(x)$, $p_{n-1}(x)$ is the orthonormal polynomial of degree n-1 for $d\hat{\beta}(x)$, and γ_{n-1} and γ_n are the leading coefficients of $p_{n-1}(x)$ and $p_{np}(x)$ respectively. Now, by orthonormality,

$$\gamma_{n-1}/\gamma_n = \int_{-\infty}^{\infty} x\, p_{n-1}(x)\, p_{np}(W,x)\, d\hat{\beta}(x)$$

$$\leq \{ \int_{-\infty}^{\infty} \{x\, p_{np}(W,x)\}^2 \, d\hat{\beta}(x) \}^{1/2}$$

$$= \{ \int_{-\infty}^{\infty} x^2 \, |p_{np}(W,x)|^p \, W(x)^p \, dx \}^{1/2},$$

by the Cauchy-Schwarz inequality. Then by Hölder's inequality,

$$(8.16) \qquad \gamma_{n-1}/\gamma_n \leq \|xp_{np}(W,x)W(x)\|_{L_p(\mathbb{R})} \, \|p_{np}(W,x)W(x)\|_{L_p(\mathbb{R})}^{(p-2)/(2p)}$$

$$= \|xp_{np}(W,x)W(x)\|_{L_p(\mathbb{R})}$$

$$\leq 2\, \|xp_{np}(W,x)W(x)\|_{L_p[-2a_n,2a_n]} \leq 4a_n,$$

uniformly for $2 \leq p < \infty$, by Lemma 7.4.

Next, choose an $H(x)$ as in Lemma 8.3, satisfying

$$H(x) \sim W^{-2}(x)\,(1 + x^2)^{-2}, \; x \in \mathbb{R}.$$

We have for any $P \in \mathcal{P}_{n-1}$,

$$|P(x)| = |\sum_{j=1}^{n} l_{jn}(x)\, P(x_{jn}^{(p)})|$$

$$\leq \gamma_{n-1}/\gamma_n \sum_{j=1}^{n} \lambda_{jn} \, |p_{n-1}(x_{jn}^{(p)})| \, |P(x_{jn}^{(p)})\, p_{np}(x)/(x - x_{jn}^{(p)})|$$

$$\leq 4a_n \max_{1\leq j\leq n} |P(x_{jn}^{(p)})\, H(x_{jn}^{(p)})^{-1/2}\, p_{np}(x)\,/(x - x_{jn}^{(p)})|$$

$$\times \{ \sum_{j=1}^{n} \lambda_{jn}\, p_{n-1}^2(x_{jn}^{(p)})\}^{1/2} \{ \sum_{j=1}^{n} \lambda_{jn}\, H(x_{jn}^{(p)}) \}^{1/2}$$

$$\leq 4a_n\, C_5\, (1 + (x_{1n}^{(p)})^2) \max_{1\leq j\leq n} |PW(x_{jn}^{(p)})\, p_{np}(x)/(x - x_{jn}^{(p)})|$$

$$\times \{ \int_{-\infty}^{\infty} H(x)\, d\hat{\beta}(x) \}^{1/2},$$

by the Cauchy-Schwarz inequality, and the Gauss-quadrature formula, and by Lemma 8.4 and choice of $H(x)$. Now

$$\int_{-\infty}^{\infty} H(x)\, d\hat{\beta}(x) \leq C_8 \int_{-\infty}^{\infty} (1 + x^2)^{-2}\, |p_{np}(x)|^{p-2}\, W(x)^{p-2}\, dx$$

$$\leq C_8 \, \|(1+x^2)^{-1}\|^2_{L_p(\mathbb{R})} \, \|p_{np}(x)W(x)\|^{1-2/p}_{L_p(\mathbb{R})} \leq 4C_8,$$

where we have used Hölder's inequality. Here C_8 depends only on H and W, and so is independent of n and p. Using (8.12), and the uniform in p upper bound on $x^{(p)}_{1n}$ that we have already established, we obtain (8.14) for $2 \leq p < \infty$, with constants independent of n and p. To obtain (8.15), we may use the Nikolskii inequality in Lemma 7.3, which shows that

$$\|p_{np}W\|_{L_\infty(\mathbb{R})} \leq \{W(0)/W(2)\} \, (C\, n^2),$$

the constant C being independent of n and p.

Finally, to obtain the results for $p = \infty$, one uses the facts that the constants are independent of p, and that we have uniformly in compact subsets of $\mathbb{C}$, as $p \to \infty$,

$$T_{np}(W,x) \to T_{n\infty}(W,x),$$
$$p_{np}(W,x) \to p_{n\infty}(W,x),$$

and so

$$x^{(p)}_{jn} \to x^{(\infty)}_{jn}, \quad j = 1,2,\ldots,n.$$

These limiting relations are well known, and may be proved using the compactness of P_n with norm $\|P\| := \|PW\|_{L_\infty(\mathbb{R})}$, as well as the uniqueness of $T_{n\infty}(W,x)$. □

Proof of Theorem 8.1(a)

We use the polynomials of Lemma 6.3, and Lemma 8.6 to obtain a contradiction if the conclusion of the theorem fails to be true. Let us suppose that for some fixed positive integer j, we have proved for $2 \leq p \leq \infty$ and $n \geq n_1$,

$$x^{(p)}_{j-1,n}/a_n - 1 \geq -C_3\epsilon_n,$$

where

$$\epsilon_n := ((\log n)/n)^{2/3}.$$

In the case $j = 1$, this hypothesis is vacuous. Setting

$$P(x) := S(x) \prod_{k=1}^{j-1} (x^2 - (x_{kn}^{(p)})^2),$$

in (8.15), where $S \in P_{n-2j}$, we obtain for $|x| < x_{j-1,n}^{(p)}$,

$$|S(x)W(x)| \leq C_3 n^{C_4} (2(x_{1n}^{(p)})^2)^{j-1} ((x_{j-1,n}^{(p)})^2 - x^2)^{-(j-1)}$$

$$\times \max_{j \leq k \leq n+1-j} |SW(x_{kn}^{(p)}) / (x - x_{kn}^{(p)})|,$$

since the zeros of $T_{np}(W,x)$ are symmetric about 0. Here, if $j = 1$, we interpret $x_{j-1,n}^{(p)} = x_{0n}^{(p)}$ as a_n. Let us assume now that for some $K \geq 2$,

$$x_{jn}^{(p)} \leq x_{j-1,n}^{(p)} (1 - 3K\epsilon_n).$$

Then we obtain, using our upper bound in Theorem 8.1(b), that

$$(8.17) \qquad |S(x)W(x)| \leq C_5 n^{C_6} \|SW\|_{L_\infty(|x| \leq x_{jn}^{(p)})},$$

for $x \in I := (x_{j-1,n}^{(p)}(1 - 2K\epsilon_n), x_{j-1,n}^{(p)}(1 - K\epsilon_n))$. Here C_5, C_6 are independent of x, S, n, p and K, but depend on j and W.

We now use Lemma 6.3 to derive a contradiction if K is too large. Let

$$R := x_{j-1,n}^{(p)}(1 - 2K\epsilon_n),$$

and

$$S(x) := \hat{P}_{n-2j}(x/R) \in P_{n-2j},$$

where $\hat{P}_{n-2j}(x)$ is the polynomial satisfying (6.32) and (6.33) in Lemma 6.3 with n replaced by $n-2j$. Since $x_{jn}^{(p)} \leq R$, (6.32) shows that

$$\|SW\|_{L_\infty(|x| \leq x_{jn}^{(p)})} \leq \|\hat{P}_{n-2j}(x)W(Rx)\|_{L_\infty[-1,1]} \leq C_7 n^{C_8}.$$

Also by (6.33), for $1 \leq |x|/R \leq 2$,

$$|S(x)W(x)| \geq C_9 \exp((n-2j)\, U_{n-2j,R}(x/R)).$$

Thus, from (8.17), for $n \geq n_1(K)$,

$$(8.18) \qquad \exp((n-2j)\, U_{n-2j,R}(x/R)) \leq C_{10} n^{C_{11}},$$

for $x \in I$, since $x \in I$ implies that

$$1 \le x/R \le (1 - K\epsilon_n) / (1 - 2K\epsilon_n) \le 2$$

provided $\epsilon_n K \le 1/3$. Now by (6.29) if $x/R := 1 + \delta$,

$$U_{n-2j}(x/R) \ge -C_{12}\delta^{3/2} + C_{13}(1 - R/a_{n-2j})\,\delta^{1/2}.$$

Here by our upper bound in Theorem 8.1(b) and Lemma 8.5,

$$\begin{aligned} R/a_{n-2j} &= (x^{(p)}_{j-1,n}/a_n)\,(a_n/a_{n-2j})\,(1 - 2K\epsilon_n) \\ &\le (1 + C_{14}\epsilon_n)\,(1 + C_{15}j/n)\,(1 - 2K\epsilon_n) \\ &\le 1 - K\epsilon_n, \end{aligned}$$

provided that K is large enough, since C_{14}, C_{15} are independent of K (and also, incidentally of p). Let us set

$$\delta := K^{1/4}\epsilon_n.$$

Then

$$\begin{aligned} (n - 2j)U_{n-2j}(x/R) &\ge (n - 2j)(-C_{12}K^{3/4} + C_{13}K^{5/4})\,(\log n)/n \\ &\ge C_{15}K^{5/4}\log n, \end{aligned}$$

if K is large enough, the lower bound on K, being independent of p. This last relation will contradict (8.18) if K is large enough and $n \ge n_1(K)$, provided that $x \in I$, that is, provided

$$R(1 + \delta) \le x^{(p)}_{j-1,n}(1 - K\epsilon_n).$$

This last inequality is easy to verify from the definition of R, δ and the fact that $K > 1$. Thus for some fixed K, and $n \ge n_1(K)$, both n_1 and K being independent of p,

$$\begin{aligned} x^{(p)}_{jn} &> x^{(p)}_{j-1,n}(1 - 3K\epsilon_n) \\ &\ge a_n(1 - C_{16}((\log n)/n)^{2/3}). \end{aligned}$$

Since all the constants are independent of $2 \le p < \infty$, we can let $p \to \infty$ to obtain the inequality for $p = \infty$ as well. □

<u>Proof of Corollary 8.2</u>

The inequality $y_{1n} \le 1$ follows from (7.4). The lower bounds in (8.4) follow from the interlacing of zeros and extrema of $T_{n\infty}(W,x)W(x)$. □

9. Further Properties of $U_{n,R}(x)$.

In Lemma 5.3, we established various properties of $U_{n,R}(x)$ defined by (5.51). Under additional assumptions, further properties were established in Lemma 6.2, and in Theorem 7.1(ii), $U_{n,R}(x)$ was shown to have an important role in bounding the growth of weighted polynomials in the plane. In this section, we prove that $U_{n,R}(z)$ is positive in a certain region containing $(-1,1)$, that is independent of n and R if $a_n/2 \le R \le a_n$:

Theorem 9.1

Let $W(x) := e^{-Q(x)}$, where $Q(x)$ is even and continuous in $\mathbb{R}$, $Q''(x)$ exists and is continuous in $(0,\infty)$ and $Q'(x)$ is positive in $(0,\infty)$, while for some $C_1, C_2 > 0$,

$$(9.1)\qquad C_1 \le (xQ'(x))'/Q'(x) \le C_2,\ x \in (0,\infty).$$

For $0 < \theta < \pi/2$ and $\epsilon > 0$, let

$$(9.2)\qquad G(\epsilon;\theta) := \{z\colon |\mathrm{Re}(z)| \le 1;\ 0 < |\mathrm{Im}(z)| < \epsilon\}$$

$$\cup\ \{\pm z\colon 0 < |\mathrm{Im}(z)| < \epsilon\ ;\theta < |\arg(z-1)| \le \pi/2\}.$$

Let $0 < \delta < \pi/6$.

Then there exist C and $\epsilon_0 > 0$ such that for $n \ge 2$ and $a_n/2 \le R \le a_n$, and $x+iy \in G(\epsilon_0;(\pi/3)+\delta)$ or $0 < |x+iy| \le 1$,

$$(9.3)\qquad U_{n,R}(x+iy) \ge Cy^2/(x^2+y^2)^{1/2},$$

with C and ϵ_0 depending on δ, but independent of n, R, x and y.

Note that $G(\epsilon;\theta)$ is an open set not intersecting the real axis (see Figure 9.1). For the case $Q(x) := |x|^\alpha$, $\alpha > 0$, the above theorem is essentially Theorem 3.9 in [32], proved in the appendix of [32]. Here we use some of the ideas in [32], slightly simplifying the proofs, but obtaining weaker conclusions.

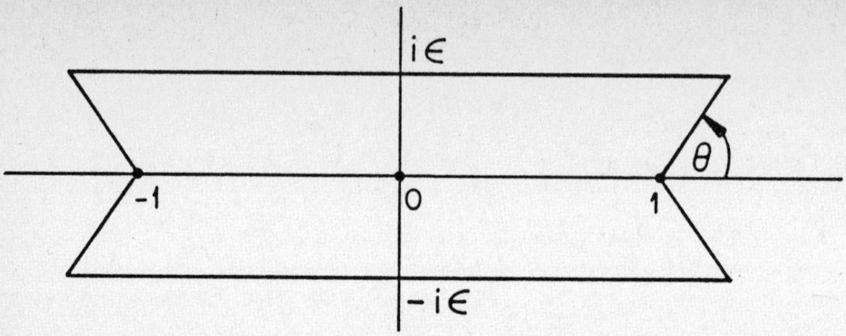

Figure 9.1 : $G(\epsilon;\theta)$

Lemma 9.2

Let $W(x) := e^{-Q(x)}$ be as in Theorem 9.1. Then for $n \geq 2$, $a_n/2 \leq R \leq a_n$ and

(9.4) $\quad \Delta := (x^2 + y^2)^{1/2} \leq 1,$

we have

(9.5) $\quad U_{n,R}(x + iy) \geq Cy^2/\Delta,$

where C is independent of n and R.

Proof

Recall from (5.45) and (5.52) that since $z := x+iy$ has $|z| = \Delta \leq 1$,

$$Q(R|z|)/n = Q(R\Delta)/n$$

$$= \int_{-1}^{1} \log |\Delta - t| \, \mu_{n,R}(t) \, dt + \chi_{n,R}/n.$$

Then by the definition (5.51) of $U_{n,R}(z)$,

$$U_{n,R}(z) = \int_{-1}^{1} \log |z - t| \, \mu_{n,R}(t) \, dt - \int_{-1}^{1} \log |\Delta - t| \, \mu_{n,R}(t) \, dt$$

$$= (1/2) \int_{-1}^{1} \log \left|\frac{(x - t)^2 + y^2}{(\Delta - t)^2}\right| \mu_{n,R}(t) \, dt$$

$$= (1/2) \int_{-1}^{1} \log \left|1 + \frac{2t \, (\Delta - x)}{(\Delta - t)^2}\right| \mu_{n,R}(t) \, dt,$$

by definition of Δ. As $\mu_{n,R}(t)$ is even (see (5.37) and (5.38) and recall that $tQ'(t)$ is even), we obtain

$$U_{n,R}(z) = (1/2)\int_0^1 \log\{|1 + \frac{2t\ (\Delta - x)}{(\Delta - t)^2}||1 - \frac{2t\ (\Delta - x)}{(\Delta + t)^2}|\}\ \mu_{n,R}(t)dt,$$

$$= (1/2)\int_0^1 \log\{\ 1 + \frac{4\ t^2 y^2}{(\Delta^2 - t^2)^2}\ \}\ \mu_{n,R}(t)\ dt,$$

after some elementary manipulations, noting that $\Delta^2 - x^2 = y^2$. Next, noting that

$$(\Delta^2 - t^2)^2 \le \Delta^4,\ t \in (0,\Delta),$$

and

$$\log\ (1 + 4u) \ge u/2,\ u \in (0,1),$$

we obtain

$$U_{n,R}(z) \ge (1/2)\int_0^{\Delta} \log\{\ 1 + 4t^2y^2\Delta^{-4}\ \}\ \mu_{n,R}(t)\ dt, \tag{9.6}$$

$$\ge (y^2/(4\ \Delta^4))\int_0^{\Delta} t^2\ \mu_{n,R}(t)\ dt$$

$$\ge (y^2/(12\Delta))\ \min\{\ \mu_{n,R}(t):\ t \in [0,1/2]\ \},$$

as $\Delta/2 \le 1/2$. Next, for $t \in [0,1/2]$, (5.38) and (5.40) show that

$$\mu_{n,R}(t) \ge 2\pi^{-2}\sqrt{1 - t^2}\int_{3/4}^{1} \frac{RsQ'(Rs) - RtQ'(Rt)}{n(s^2 - t^2)(1 - s^2)^{1/2}}\ ds$$

$$\ge 2\pi^{-2}\sqrt{3/4}\ ((3R/4)Q'(3R/4)-(R/2)Q'(R/2))(16/5)\ n^{-1}\ \arccos(3/4)$$

$$= C_3\ R\ (4n)^{-1}\ (uQ'(u))',$$

where C_3 is an absolute constant, and u lies between R/2 and 3R/4. By (9.1) and monotonicity of $tQ'(t)$, we obtain for $t \in [0,1/2]$, $a_n/2 \le R \le a_n$,

$$\mu_{n,R}(t) \ge C_4\ (a_n/4)\ Q'(a_n/4)\ /n \ge C_5,$$

by (6.19) and (6.21). Then (9.6) yields (9.5). □

The next lemma follows the proof of Lemma A.2 in [32]:

<u>Lemma 9.3</u>

<u>Let</u> $W(x) := e^{-Q(x)}$ <u>be as in Theorem 9.1. Let</u> $0 < \delta < \pi/6$. <u>Then there exist</u> $\epsilon_0 > 0$ <u>and</u> C <u>such that for</u> $n \ge 2$ <u>and</u> $a_n/2 \le R \le a_n$,

$$(9.7) \qquad U_{n,R}(1 + ye^{i\theta}) \geq Cy^{3/2}, \ y \in [0,\epsilon_0], \theta \in [\pi/3 + \delta, 3\pi/4],$$

<u>where</u> ϵ_0 <u>and</u> C <u>are independent of</u> n,R,y <u>and</u> θ.

<u>Proof</u>

By (5.29),(5.37) and (5.51),

$$U_{n,R}(z) = \int_{-1}^{1} \log|z - t|\, v_{n,R}(t)\,dt + B_{n,R} \log|\varphi(z)/2|$$
$$- Q(R|z|)/n + \chi_{n,R}/n.$$

Setting $z := 1 + ye^{i\theta}$, and noting that $|\varphi(1 + ye^{i\theta})|$ increases with increasing y, we have

$$(9.8) \qquad \frac{\partial U_{n,R}(z)}{\partial y} \geq \int_{-1}^{1} \frac{(1-t)\cos\theta + y}{|z - t|^2} v_{n,R}(t)dt - RQ'(R|z|)(y+\cos\theta)/(n|z|).$$

Here for some u between R and R|z|,

$$R|z|Q'(R|z|) - RQ'(R) = (R|z| - R)\ (uQ'(u))'$$
$$= O(R\ ||z| - 1|\ Q'(u))$$
$$= O(y\ 2RQ'(2R))$$

by monotonicity of uQ'(u), uniformly for $|y| \leq 1/2$, $|\theta| \leq \pi$. Then

$$RQ'(R|z|)/|z| = RQ'(R)/|z|^2 + O(y\ 2RQ'(2R)/|z|)$$
$$= RQ'(R) + O(y\ 2RQ'(2R))$$
$$= RQ'(R) + O(ny),$$

uniformly for $|y| \leq 1/2$, $|\theta| \leq \pi$, $R \leq a_n$, by (6.19) and (6.21). Then using (5.50), we have

$$(9.9) \qquad \frac{\partial U_{n,R}(z)}{\partial y} \geq \int_{-1}^{1} \left\{\frac{(1-t)\cos\theta + y}{|z - t|^2} - \frac{\cos\theta}{1 - t}\right\} v_{n,R}(t)\,dt - C_5 y,$$

where C_5 is independent of R,n,y and θ. Here, by the substitution $1-t=uy$,

$$(9.10) \qquad I := \int_{-1}^{1} \left\{\frac{(1-t)\cos\theta + y}{|z - t|^2} - \frac{\cos\theta}{1 - t}\right\} v_{n,R}(t)\,dt$$
$$= \int_{0}^{2/y} \left\{\frac{u\cos\theta + 1}{|u + e^{i\theta}|^2} - \frac{\cos\theta}{u}\right\} v_{n,R}(1 - uy)\,du$$

$$= \int_0^{2/y} \frac{1 - 2\cos^2\theta - \cos\theta/u}{|u + e^{i\theta}|^2} \upsilon_{n,R}(1 - uy)\, du.$$

Here, if $\theta \in [\pi/2, 3\pi/4]$,

$$1 - 2\cos^2\theta - \cos\theta/u \geq 1 - 2\cos^2\theta \geq 0,$$

while if $\theta \in [\pi/3, \pi/2]$ and $y^{-1/2} \leq u \leq 2/y$,

$$1 - 2\cos^2\theta - \cos\theta/u \geq 1/2 - y^{1/2} \geq 0,$$

if $y \leq 1/4$. Hence, for $0 \leq y \leq 1/4$, $\theta \in [\pi/3, 3\pi/4]$,

$$I \geq \int_0^{y^{-1/2}} \frac{1 - 2\cos^2\theta - \cos\theta/u}{|u + e^{i\theta}|^2} \upsilon_{n,R}(1 - uy)\, du.$$

Next, by (5.49) and (6.26), we have uniformly for $a_n/2 \leq R \leq a_n$, $n \geq 1$, $0 \leq u \leq y^{-1/2}$ and $y^{1/2} \leq 1/8$,

$$\upsilon_{n,R}(1 - uy) = A_{n,R}\, (1 - (1 - uy)^2)^{1/2} + O(1 - (1 - uy)^2)^{7/10}$$

$$= A_{n,R}\, (2uy)^{1/2}\, (1 + O(y^{1/2})) + O\, ((uy)^{7/10}).$$

Then

$$I \geq A_{n,R}(2y)^{1/2} \int_0^{y^{-1/2}} \frac{1 - 2\cos^2\theta - \cos\theta/u}{|u + e^{i\theta}|^2} u^{1/2}\, du$$

$$+ O(A_{n,R} y \int_0^{y^{-1/2}} \frac{(1 + u^{-1})}{|u + e^{i\theta}|^2} u^{1/2}\, du)$$

$$+ O(y^{7/10} \int_0^{y^{-1/2}} \frac{(1 + u^{-1})}{|u + e^{i\theta}|^2} u^{7/10}\, du).$$

Using (6.24), and the fact that for $\theta \in [\pi/3, 3\pi/4]$,

$$|u + e^{i\theta}|^2 \geq u^2 - \sqrt{2}u + 1 \geq (1 - 1/\sqrt{2})\, (u^2 + 1),\ u \geq 0,$$

we obtain

$$I \geq A_{n,R}(2y)^{1/2}\Big[\int_0^{\infty} \frac{1 - 2\cos^2\theta - \cos\theta/u}{|u + e^{i\theta}|^2} u^{1/2} du + O(y^{1/4})\Big] + O(y^{7/10}).$$

Here, by the substitution $u = v^{-1}$,

$$g(\theta) := \int_0^{\infty} |u + e^{i\theta}|^{-2}\, u^{1/2}\, du = \int_0^{\infty} |v + e^{i\theta}|^{-2}\, v^{-1/2}\, dv,$$

so by (6.24) again,

$$I \geq C_6\, y^{1/2}\, (1 - 2\cos^2\theta - \cos\theta)\, g(\theta) + O(y^{7/10}),$$

where C_6 is independent of n, y, θ. Now if $\theta \in [\pi/3 + \delta, 3\pi/4]$,

$$g(\theta) \geq C_7 \quad \text{and} \quad 1 - 2\cos^2\theta - \cos\theta \geq C_8,$$

so from (9.9) and (9.10),

$$\frac{\partial U_{n,R}(z)}{\partial y} \geq C_9 y^{1/2},$$

uniformly for $n \geq 2$, $a_n/2 \leq R \leq a_n$, $0 < y \leq \epsilon_1$, $\theta \in [\pi/3+\delta, 3\pi/4]$, where ϵ_1, C_9 are independent of n, R, θ and y. Integrating and using $U_{n,R}(1) = 0$, we obtain (9.7). □

Proof of Theorem 9.1

This follows easily from Lemmas 9.2 and 9.3 and the fact that

$$U_{n,R}(\bar{z}) = U_{n,R}(-z) = U_{n,R}(z). \ \square$$

We remark that the condition $a_n/2 \leq R \leq a_n$ in Theorem 9.1 can be replaced by $a_n\eta \leq R \leq a_n$, for any fixed $\eta \in (0,1)$.

10. Nth Root Asymptotics for Extremal Polynomials.

In this section, we establish nth root asymptotics for extremal polynomials, which will be used in the next section in estimating the approximation power of certain weighted polynomials. Under various conditions on the weight, nth root asymptotics have been obtained by Rahmanov [62], Mhaskar and Saff [46], Gončar and Rahmanov [17], Luo and Nuttall [35], and Lubinsky and Saff[32]. Unfortunately, none of these results are directly applicable in the present context, since they apply when the zero distribution is an Ullman distribution of fixed order α, for some fixed $\alpha > 0$. By contrast, in the situation treated here, different subsequences of $\{p_{np}(W,x)\}_0^\infty$ may have different asymptotic zero distributions. Nevertheless, the methods of [46,62] apply with only minor changes to yield:

Theorem 10.1

Let $W(x) := e^{-Q(x)}$, where $Q(x)$ is even and continuous in $\mathbb{R}$, $Q''(x)$ exists and is continuous in $(0,\infty)$, $Q'(x)$ is positive in $(0,\infty)$, while for some $C_1, C_2 > 0$,

$$(10.1)\qquad C_1 \le (xQ'(x))'/Q'(x) \le C_2, \quad x \in (0,\infty).$$

For $n = 1,2,3,\ldots$, let $0 < R_n \le a_n$, and $S_n(x)$ be a monic polynomial of degree n satisfying

$$(10.2)\qquad \|S_n W\|_{L_\infty[-R_n,R_n]} = E_n := \min_{P \in \mathcal{P}_{n-1}} \|(x^n - P(x))W(x)\|_{L_\infty[-R_n,R_n]}.$$

Then, locally uniformly in $\mathbb{C}\setminus[-1,1]$,

$$(10.3)\qquad \lim_{n\to\infty} R_n^{-1} |S_n(R_n z)|^{1/n} \exp\left(-\int_{-1}^{1} \log |z - t| \, \mu_{n,R_n}(t)\, dt\right) = 1,$$

and

$$(10.4)\qquad \lim_{n\to\infty} \{|S_n(R_n z)W(R_n|z|)| / E_n\}^{1/n} \exp(-U_{n,R_n}(z)) = 1,$$

where $\mu_{n,R}$ and $U_{n,R}$ are as in Lemma 5.3. Furthermore if $\chi_{n,R}$ is given by (5.52),

$$(10.5)\qquad \lim_{n\to\infty} E_n^{1/n}\, R_n^{-1} \exp(\chi_{n,R_n}/n) = 1.$$

As a corollary, we have:

Corollary 10.2

Let $W(x)$ be as in Theorem 10.1. Then locally uniformly in $\mathbb{C}\backslash[-1,1]$,

$$(10.6)\qquad \lim_{n\to\infty} a_n^{-1}\, |T_{n\infty}(W,a_n z)|^{1/n} \exp\left(-\int_{-1}^{1} \log|z-t|\, \mu_{n,a_n}(t)dt\right) = 1,$$

and

$$(10.7)\qquad \lim_{n\to\infty} \{|T_{n\infty}(W,a_n z)W(a_n|z|)| \,/\, E_{n\infty}(W)\}^{1/n} \exp(-U_{n,a_n}(z)) = 1.$$

Further, if the zeros of $T_{n\infty}(W,a_n z)$ are

$$(10.8)\qquad -\infty < x_{nn}^{(\infty)} < x_{n-1,n}^{(\infty)} < \ldots < x_{1n}^{(\infty)} < \infty,$$

then for each function $f(x)$ bounded and piecewise continuous in $[-1,1]$,

$$(10.9)\qquad \lim_{n\to\infty} \left\{ n^{-1} \sum_{j=1}^{n} f(x_{jn}^{(\infty)}/a_n) - \int_{-1}^{1} f(t)\, \mu_{n,a_n}(t)\, dt\right\} = 0.$$

It is possible to give an L_p analogue of Theorem 10.1 and Corollary 10.2, and to prove results for weights $\hat{W}(x)$ for which $\log \hat{W}(x)/Q(x) \to -1$ as $|x| \to \infty$. One technical consequence of Corollary 10.2 and the results of Section 8 is:

Corollary 10.3

Assume the notation and hypotheses of Corollary 10.2. Let $0 < \delta < \pi/6$. Then there exists $\epsilon_0 > 0$ such that for each compact subset K of $G(\epsilon_0;(\pi/3)+\delta)$,

$$(10.10)\qquad \limsup_{n\to\infty} \{ E_{n\infty}(W) \,/\, \min_{z\in K} |T_{n\infty}(W,a_n z)W(a_n|z|)| \}^{1/n} < 1.$$

Here $G(\epsilon_0;(\pi/3)+\delta)$ is defined by (9.2). Furthermore, if l is a fixed positive integer and

(10.11) $\xi_n := y_{1n}/a_n$,

<u>where</u> y_{1n} <u>is the largest equioscillation point of</u> $T_{n\infty}(W,x)W(x)$ <u>and</u>

(10.12) $z := (\xi_n^2 + iy)^{1/2}, \quad 0 < y \le \epsilon_0$,

<u>then</u>

(10.13) $E_{n\infty}(W)/|T_{n\infty}(W,a_n z)W(a_n|z|)| \le (1 + Cy^2((\log n)/n)^{-4/3})^{-l}$,

<u>where</u> C <u>is independent of</u> n <u>and</u> y.

<u>Proof of Theorem 10.1</u>

We first prove (10.5). Now if $P(x)$ is any monic polynomial of degree n, (7.5) shows that for $z \in \mathbb{C}\setminus[-1,1]$, $0 < R \le a_n$,

$$|P(z)W(R|z|)\exp(-nU_{n,R}(z))| \le \|P(x)W(Rx)\|_{L_\infty[-1,1]}.$$

Letting $z \to \infty$, and taking account of the definition (5.51) of $U_{n,R}(z)$, we obtain

$$\exp(-\chi_{n,R_n}) \le \|P(x)W(Rx)\|_{L_\infty[-1,1]}.$$

Thus

(10.14) $$\hat{E}_n := \min_{\hat{P} \in P_{n-1}} \|(x^n - \hat{P}(x))W(R_n x)\|_{L_\infty[-1,1]} \ge \exp(-\chi_{n,R_n}).$$

To obtain an upper bound, we use Lemma 6.3. Let $\hat{P}_n(x)$ denote the polynomial with leading coefficient $\exp(\chi_{n,R_n})$ satisfying (6.32). By (6.32),

$$\hat{E}_n \le \| \{\hat{P}_n(x)\exp(-\chi_{n,R_n})\} W(R_n x) \|_{L_\infty[-1,1]}$$
$$\le \exp(-\chi_{n,R_n})\, C_4 n^{C_5},$$

with C_4, C_5 independent of n. Together with (10.14), this yields

$$\lim_{n\to\infty} \hat{E}_n^{1/n} \exp(\chi_{n,R_n}/n) = 1.$$

Since, by a substitution $x = R_n u$ in (10.2), $E_n = R_n^n \hat{E}_n$, we obtain (10.5). Next, for $z \in \mathbb{C}\setminus[-1,1]$, (7.5) shows that

$$(10.15)\quad |S_n(R_nz)W(R_n|z|)|/E_n \le \exp(nU_{n,R_n}(z))\|S_n(R_nx)W(R_nx)\|_{L_\infty[-1,1]}/E_n$$

$$= \exp(nU_{n,R_n}(z)),$$

by definition of E_n. Let

$$f_n(z) := n^{-1} \log \{|S_n(R_nz)W(R_n|z|)| \exp(-nU_{n,R_n}(z))/E_n\}$$

$$= n^{-1} \log |R_n^{-n} S_n(R_nz)| - \int_{-1}^{1} \log |z - t| \mu_{n,R_n}(t)\, dt$$

$$-\chi_{n,R_n}/n - (\log E_n)/n + \log R_n,$$

a function harmonic in $\overline{\mathbb{C}}\setminus[-1,1]$. By (10.15),

$$f_n(z) \le 0,\ z \in \mathbb{C}\setminus[-1,1],\ n \ge 1,$$

while from (10.5),

$$f_n(\infty) = -\chi_{n,R_n}/n - (\log E_n)/n + \log R_n \to 0,\ n \to \infty.$$

It follows from the maximum principle for harmonic functions that

$$\lim_{n\to\infty} f_n(z) = 0,$$

locally uniformly in $\mathbb{C}\setminus[-1,1]$. Then (10.3) and (10.4) follow. □

Proof of Corollary 10.2

Choosing $R_n = a_n$, we see from (7.3) that $S_n(z) := T_{n\infty}(W,z)$ satisfies (10.2). Then (10.6) and (10.7) follow from Theorem 10.1. The quadrature convergence (10.9) then follows by a standard argument - see, for example, [46,pp.226-7]. □

Proof of Corollary 10.3

Let $0 < \delta < \pi/6$ and ϵ_0 be the associated number in Theorem 9.1. By (10.7), if K is a compact subset of $G(\epsilon_0;(\pi/3)+\delta)$,

$$\limsup_{n\to\infty} \{ E_{n\infty}(W) / \min_{z\in K} |T_{n\infty}(W,a_nz)W(a_n|z|)| \}^{1/n}$$

$$\le \exp(-\liminf_{n\to\infty} \{ \min_{z\in K} U_{n,a_n}(z) \}) < 1,$$

by (9.3). The proof of (10.13) is somewhat harder. Let

$$z_{jn} := x_{jn}^{(\infty)}/a_n, \quad j=1,2,\dots n.$$

Since $W(x)$ is even, the zeros $x_{jn}^{(\infty)}$ are symmetric about 0. Then with ξ_n and z as in (10.11) and (10.12),

$$(10.16) \quad E_{n\infty}(W) \,/\, |T_{n\infty}(W,a_n z)W(a_n|z|)|$$

$$= |T_{n\infty}(W,a_n\xi_n) \,/\, T_{n\infty}(W,a_n z)| \; W(a_n\xi_n) \,/\, W(a_n|z|)$$

$$= \{ \prod_{z_{jn}>0} |\xi_n^2 - z_{jn}^2| \,/\, |z^2 - z_{jn}^2| \; |\xi_n/z|^{n-2\langle n/2\rangle} \} \; W(a_n\xi_n)/W(a_n|z|)$$

$$= \{ \prod_{z_{jn}>0} |1 + iy/(\xi_n^2 - z_{jn}^2)|^{-1} \} |1+iy/\xi_n^2|^{\langle n/2\rangle - n/2} W(a_n\xi_n)/W(a_n|z|).$$

Here $\langle x\rangle$ denotes the greatest integer $\le x$. Now by Corollary 8.2,

$$(10.17) \quad 1 - C((\log n)/n)^{2/3} \le \xi_n \le 1,$$

for n large enough. Then

$$|z| = (\xi_n^4 + y^2)^{1/4} \le \xi_n + 2y^2,$$

for n large enough, $0 < \epsilon \le \epsilon_0$, by the inequality

$$(a^4 + b)^{1/4} \le a + 2b, \; b \ge 0, \; a \ge 1/2.$$

Hence

$$Q(a_n|z|) - Q(a_n\xi_n) \le Q(a_n(\xi_n + 2y^2)) - Q(a_n\xi_n) = a_n 2y^2 Q'(a_n c),$$

for some c between ξ_n and $\xi_n + 2y^2$. Using the monotonicity of $uQ'(u)$, we obtain for n large enough and $0 < y \le \epsilon_0$,

$$Q(a_n|z|) - Q(a_n\xi_n) \le 5a_n Q'(2a_n)y^2 \le C_3\, ny^2,$$

by (6.19) and (6.21). Thus

$$(10.18) \quad W(a_n\xi_n) \,/\, W(a_n|z|) \le \exp(\, C_3\, ny^2), \quad 0 < y \le \epsilon_0.$$

Next, using Theorem 8.1 and (10.17), we obtain

$$\xi_n^2 - z_{jn}^2 \le C((\log n)/n)^{2/3}, \quad j=1,2,\dots,2l.$$

Then

$$(10.19) \quad \prod_{j=1}^{2l} |1 + iy/(\xi_n^2 - z_{jn}^2)|^{-1} = \prod_{j=1}^{2l} \{1 + y^2/(\xi_n^2 - z_{jn}^2)^2\}^{-1/2}$$

$$\le \{1 + C_5 y^2((\log n)/n)^{-4/3}\}^{-l}, \quad 0 < y \le \epsilon_0,$$

n large enough. Next, if $0 < r < 1$, we have

$$S := \prod_{j=2l+1}^{\langle n/2\rangle} |1 + iy/(\xi_n^2 - z_{jn}^2)^2|^{-1}$$

$$= \exp\Big(-\tfrac{1}{2} \sum_{j=2l+1}^{\langle n/2\rangle} \log \{ 1 + y^2/(\xi_n^2 - z_{jn}^2)^2 \} \Big)$$

$$\le \exp\Big(-\tfrac{1}{2} \sum_{0<z_{jn}<1-r} \log \{ 1 + y^2/(\xi_n^2 - z_{jn}^2)^2 \} \Big)$$

for $n \ge n_1(r)$, independent of y, by Theorem 8.1. Now if $y \le r$, $0 < z_{jn} < 1-r$, and n is large,

$$y^2 / (\xi_n^2 - z_{jn}^2)^2 \le r^2 / (\xi_n^2 - z_{jn}^2)^2 \le 1,$$

so by the inequality $\log (1 + u) \ge u/2$, $u \in [0,1]$, we have for $0 < y \le r$,

$$S \le \exp\Big(-y^2/4 \sum_{0<z_{jn}<1-r} (1 - z_{jn}^2)^{-2} \Big). \tag{10.20}$$

But by (10.9),

$$\lim_{n\to\infty} \Big\{ n^{-1} \sum_{0<z_{jn}<1-r} (1 - z_{jn}^2)^{-2} - \int_0^{1-r} (1 - t^2)^{-2} \mu_{n,a_n}(t)\, dt \Big\} = 0.$$

Here, by (5.49), for $n \ge 2$,

$$\mu_{n,a_n}(t) = v_{n,a_n}(t) \ge A_{n,a_n} (1 - t^2)^{1/2} - C_4 (1 - t^2)^{7/10} \tau_{n,a_n}$$
$$\ge C_7 (1 - t^2)^{1/2},$$

if $t \in [1-\eta,1]$, η small enough, by (6.24) and (6.26). Note that C_4 and C_7 are independent of r. Then for $r < \eta$,

$$\int_0^{1-r} (1 - t)^{-2} \mu_{n,a_n}(t)\, dt \ge C_7 \int_{1-\eta}^{1-r} (1 - t^2)^{-3/2}\, dt.$$

Choosing r small enough, we obtain for $n \ge 2$,

$$\int_0^{1-r} (1 - t^2)^{-2} \mu_{n,a_n}(t)\, dt \ge 12\, C_3,$$

where C_3 is the constant in (10.18). Then for $n \ge n_0(r)$,

$$n^{-1} \sum_{0<z_{jn}<1-r} (1 - z_{jn}^2)^{-2} \ge 8C_3,$$

so for $0 < y \le r$, $n \ge n_0(r)$, (10.20) yields

$$S \le \exp(-2n\, C_3 y^2).$$

Combined with (10.16),(10.18) and (10.19), we obtain for

$0 < y \leq \min\{\epsilon_0, r\}$, $n \geq n_0(r)$

$$E_{n\infty}(W) \,/\, |T_{n\infty}(W, a_n z) W(a_n |z|)|$$

$$\leq \{ 1 + C_5\, y^2((\log n)/n)^{-4/3} \}^{-l} \exp(-2n\, C_3 y^2 + n\, C_3 y^2),$$

giving (10.13). □

We note that the estimate (10.13) plays a crucial role in the next section. Its proof is similar to that of Lemma 4.5 in [32], though (10.13) is sharper than Lemma 4.5 in [32], as it uses the new information given by Theorem 8.1 and Corollary 8.2.

11. Approximation by Certain Weighted Polynomials, I.

In this section, we consider approximation by weighted polynomials of the form $P_n(x)W(a_nx)$, or more precisely of the form $P_n(x)/H(a_nx)$, where $H(x)$ is a certain entire function that behaves like W^{-1} on the real line. Weighted polynomial approximations of the form $P_n(x)W(a_nx)$ were first considered by Knopfmacher, Lubinsky and Nevai [21] and Lubinsky and Saff [32]. In the special case $W(x) := \exp(-|x|^\alpha)$, approximation by $P_n(x)W(a_nx)$ is equivalent to approximation by $P_n(x)W^n(x)$. The latter type of weighted polynomial had previously been considered in the framework of incomplete polynomials, and Saff [63] and Mhaskar and Saff [49] formulated certain conjectures in this connection. In the case $W(x) = \exp(-|x|^\alpha)$, the results of [32] largely resolved Saff's conjecture. For Erdös type weights, the problem has been considered by Knopfmacher and Lubinsky [20], and Lubinsky [29].

An important tool in [32] was the entire function

$$(11.1)\qquad G_Q(x) := 1 + \sum_{n=1}^{\infty} (x/q_n)^{2n}\, n^{-1/2}\, e^{2Q(q_n)},$$

where q_n is the root of $n = q_nQ'(q_n)$, for n large enough. This function was introduced ,and its behaviour as $|x| \to \infty$ was investigated in [27,28]. Rather than G_Q itself, we shall use $G_{Q/2}$. Since working with Q/2 rather than Q involves replacement of q_n by q_{2n}, we see that

$$(11.2)\qquad G_{Q/2}(x) = 1 + \sum_{n=1}^{\infty} (x/q_{2n})^{2n}\, n^{-1/2}\, e^{Q(q_{2n})}.$$

One of the two main results of this section is:

<u>Theorem 11.1</u>

<u>Let</u> $W(x) := e^{-Q(x)}$, <u>where</u> $Q(x)$ <u>is even and continuous in</u> $\mathbb{R}$, $Q''(x)$ <u>exists and is continuous in</u> $(0,\infty)$, $Q'(x)$ <u>is positive in</u> $(0,\infty)$, <u>while for some</u> $C_1, C_2 > 0$,

(11.3) $\quad C_1 \le (xQ'(x))'/Q'(x) \le C_2, \; x \in (0,\infty).$

Let $H(x)$ be an even entire function with non-negative Maclaurin series coefficients satisfying

(11.4) $\quad C_3 \le H(x)W(x) \le C_4, \; x \in \mathbb{R},$

for some $C_3, C_4 > 0$. For example, we may choose $H = G_{Q/2}$. Let $g(x)$ be a function continuous in $\mathbb{R}$, with $g(x) = 0$, $|x| \ge 1$, and let $\{k_n\}_1^\infty$ be a sequence of non-negative integers satisfying

(11.5) $\quad \lim_{n\to\infty} k_n/n = 0.$

Then there exist $P_n \in \mathcal{P}_{n-k_n}$, $n=1,2,3,\ldots$, such that

(11.6) $\quad \lim_{n\to\infty} \|g(x) - P_n(x)/H(a_n x)\|_{L_\infty(\mathbb{R})} = 0.$

The above result was proved as Theorem 4.1 in [32] when (11.3) is replaced by [32,eqn.(2.13)]

$$\lim_{|x|\to\infty} (xQ'(x))' \,/\, Q'(x) = \alpha > 0.$$

This is asymptotically a stronger condition than (11.3), forcing

$$\lim_{|x|\to\infty} \log Q(x)/\log |x| = \alpha,$$

but does not place restrictions on $Q(x)$ for small $|x|$. When we assume more about $g(x)$, we can obtain rates of convergence: Recall that $G(\varepsilon;\theta)$ is the region defined by (9.2) and illustrated in Figure 9.1.

Theorem 11.2

Let W and H be as in Theorem 11.1. Let $h(x)$ be bounded and analytic in $(-1,1) \cup G(\varepsilon_0;(\pi/3)+\delta)$ for some $\varepsilon_0 > 0$ and $0 < \delta < \pi/6$ and let $\|h\| := \sup\{|h(t)| : t \in G(\varepsilon_0;(\pi/3)+\delta)\}$. Let $\eta \ge 0$ and

(11.7)
$$g(x) := \begin{cases} h(x)\,(1-x^2)^\eta, & x \in (-1,1), \\ 0, & x \in \mathbb{R}\setminus(-1,1). \end{cases}$$

Let

(11.8) $\rho_n := ((\log n)/n)^{2/3}$, $n = 2,3,4,\ldots$,

and

(11.9) $\xi_n := y_{1n}/a_n$, $n = 2,3,4,\ldots$,

where y_{1n} is the largest point of equioscillation of $T_{n\infty}(W,x)W(x)$. Then there exists $P_n \in \mathcal{P}_n$, $n = 1,2,3,\ldots$, such that for $n = 2,3,4,\ldots$,

(11.10) $|g(x) - P_n(x)/H(a_n x)| \le C_5 \|h\| \rho_n^\eta \log(1 + \rho_n/||x| - \xi_n|)$,

$x \in \mathbb{R}$, where C_5 is independent of n,h and x. Furthermore for $n = 2,3,4,\ldots$,

(11.11) $\|g(x) - P_n(x)/H(a_n x)\|_{L_\infty(\mathbb{R})} \le C_6 \|h\| \rho_n^\eta \log n$.

It seems likely that one should be able to replace $\rho_n^\eta \log n$ in (11.11) by $n^{-2\eta/3}$, but probably nothing better. Our method of proof of Theorems 11.1 and 11.2 follows that of Theorem 4.1 in [32]. Accordingly, we first define a contour as in [32]:

Definition 11.3

Let $W(x)$ be as in Theorem 11.1, and let ξ_n be defined by (11.9). Let $0 < \epsilon < 1$ and

(11.12) $\Gamma_{n1} := \{(\xi_n^2 + iy)^{1/2} : 0 \le y \le \epsilon\}$,

for n large enough, where the branch of the square root is the principal one. Let $^{-}$ denote complex conjugation and Γ_{n2} denote the horizontal line segment joining Γ_{n1} and $(-\overline{\Gamma}_{n1})$, so that

(11.13) $\Gamma_{n2} := \{z : |\mathrm{Re}(z)| \le \mathrm{Re}(\xi_n^2+i\epsilon)^{1/2}$ and $\mathrm{Im}(z) = \mathrm{Im}(\xi_n^2+i\epsilon)^{1/2}\}$.

Finally, let

$$\Gamma_n := \Gamma_{n1} \cup \overline{\Gamma}_{n1} \cup (-\Gamma_{n1}) \cup (-\overline{\Gamma}_{n1}) \cup \Gamma_{n2} \cup \overline{\Gamma}_{n2},$$

oriented in a positive sense (see Figure 11.1).

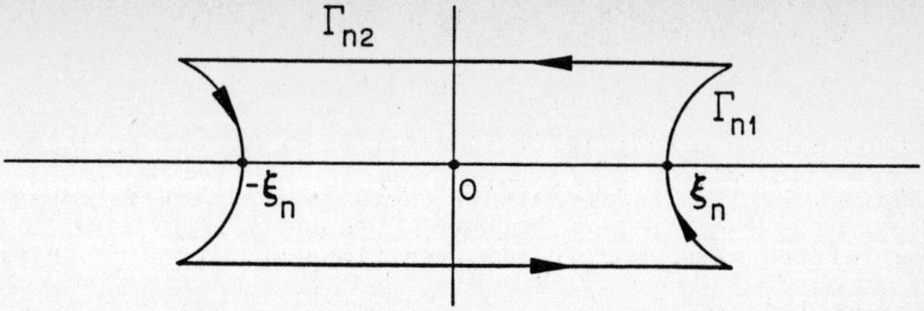

Figure 11.1 : Γ_n

Lemma 11.4

Assume the notation and hypotheses of Theorem 11.2. Let $L_n(x)$ be the Lagrange interpolation polynomial of degree at most n-1 to $H(a_n x)g(x)$ at the zeros of $T_{n\infty}(W, a_n x)$, for n = 2,3,4,... . Then for n large enough and ϵ in the definition of Γ_n small enough,

$$(11.14) \qquad |g(x) - L_n(x)/H(a_n x)| \le C \, \|h\| \, I_n(x), \quad |x| < \xi_n,$$

and

$$(11.15) \qquad |L_n(x)/H(a_n x)| \le C \, \|h\| \, I_n(x), \quad |x| > \xi_n,$$

where C is independent of n, h and x, and

$$(11.16) \qquad I_n(x) := \int_{\Gamma_{n1} \cup \Gamma_{n2}} \frac{|1 - t^2|^{\eta} \, E_{n\infty}(W)}{|t - |x|| \; |T_{n\infty}(W, a_n t) W(a_n |t|)|} \, |dt|.$$

Proof

Inside and on Γ_n, we extend g(x) to the plane by the first formula in (11.7). We show below this extension is consistent with our hypotheses on h(x). By the Hermite error formula for Lagrange interpolation,

$$(11.17) \qquad H(a_n x)g(x) - L_n(x) = \frac{1}{2\pi i} \int_{\Gamma_n} \frac{H(a_n t)g(t) \; T_{n\infty}(W, a_n x)}{(t - x) \; T_{n\infty}(W, a_n t)} \, dt,$$

for $x \in (-\xi_n, \xi_n)$. Further, the contour integral representation for $L_n(x)$ [72] shows that for $|x| > \xi_n$,

$$(11.18)\qquad -L_n(x) = \frac{1}{2\pi i}\int_{\Gamma_n} \frac{H(a_nt)g(t)\,T_{n\infty}(W,a_nx)}{(t-x)\,T_{n\infty}(W,a_nt)}\,dt.$$

Next, we show that if ϵ in (11.12) is small enough, then Γ_n is contained in $K := [-1,1] \cup \overline{G(\epsilon_0;(\pi/3)+\delta)}$, where here — denotes closure. From the definition of Γ_n and $G(\epsilon;\theta)$ (see Figures 9.1 and 11.1), we see that it suffices to show

$$\Gamma_{n1} \subset \overline{G(\epsilon_0;(\pi/3)+\delta)}.$$

So let

$$z := (\xi_n^2 + iy)^{1/2} \in \Gamma_{n1}.$$

Since by Corollary 8.2, $\xi_n \to 1$, $n \to \infty$, we have

$$z - \xi_n = iy/(2\xi_n) + y^2/(8\xi_n^3) + O(y^3),$$

uniformly for $0 < y \le \epsilon$ and $n \ge n_0(\epsilon)$. Then

$$\mathrm{Re}(z - \xi_n) \sim y^2 \ ;\ \mathrm{Im}(z - \xi_n) \sim y,$$

so

$$\arg(z - \xi_n) \ge \arctan(Cy^{-1}) \ge \arctan(C\epsilon^{-1}) \ge \pi/3+\delta,$$

if ϵ is small enough. Since $\xi_n \le 1$ (Corollary 8.2), it then also follows that $z \in G(\epsilon_0;(\pi/3)+\delta)$ for ϵ in the definition of Γ_n small enough, $0 < y \le \epsilon$ and $n \ge n_0(\epsilon)$.

Then by boundedness of h in the interior K^0 of K,

$$(11.19)\qquad |g(t)| \le \|h\|\ |1 - t^2|^{\eta},\ t \in K^0.$$

We remark that our choice of Γ_n above ensures that the only points of Γ_n possibly not in $G(\epsilon_0;(\pi/3)+\delta)$ are $\pm\xi_n$. This occurs when $\xi_n = 1$, and by continuity, (11.19) remains valid.

Next, since H is even and has non-negative Maclaurin series coefficients,

$$(11.20)\qquad \left| H(a_nt)/H(a_nx) \right| \le H(a_n|t|)/H(a_nx) \le C_5\ W(a_nx)/W(a_n|t|),\ t \in \mathbb{C},\ x \in \mathbb{R},$$

by (11.4). Dividing (11.17) by $H(a_nx)$, and using (11.19),(11.20) we obtain for $x \in (-\xi_n,\xi_n)$,

$$\begin{aligned}&|g(x) - L_n(x)/H(a_nx)| \\ &\le C_6 \ \|h\| \int_{\Gamma_n} \frac{|1 - t^2|^\eta \ |T_{n\infty}(W,a_nx)W(a_nx)|}{|t - |x|| \ |T_{n\infty}(W,a_nt)W(a_n|t|)|} \ |dt| \\ &\le 4C_6 \ \|h\| \int_{\Gamma_{n1}\cup\Gamma_{n2}} \frac{|1 - t^2|^\eta \ E_{n\infty}(W)}{|t - |x|| \ |T_{n\infty}(W,a_nt)W(a_n|t|)|} \ |dt|,\end{aligned}$$

by the symmetry of Γ_n, the evenness and reality of $T_{n\infty}(W,x)W(a_nx)$ and the reflection principle. Thus (11.14) holds. Dividing (11.18) by $H(a_nx)$, we similarly obtain (11.15). □

Note that the only place where the form of the Maclaurin series of H is used is in (11.20). We next estimate $I_n(x)$.

<u>Proof of (11.10) in Theorem 11.2</u>

First note that if ϵ in the definition of Γ_n is small enough, there exists a compact subset A of $G(\epsilon_0;\pi/3+\delta)$ containing Γ_{n2} for n large enough. Indeed Γ_{n2} is a horizontal line segment, symmetric about the imaginary axis, with right endpoint $z_n := (\xi_n^2+i\epsilon)^{1/2}$. As in the proof of Lemma 11.4, we can ensure that for n large enough,

$$\arg(z_n - \xi_n) \ge \arctan (C\epsilon^{-1}) > \pi/3 + \delta.$$

Hence the existence of the desired A. By (10.10) in Corollary 10.3, there exists $0 < \theta < 1$ such that

$$E_{n\infty}(W) \ / \min_{z\in\Gamma_{n2}} |T_{n\infty}(W,a_nz)W(a_n|z|)| \le \theta^n,$$

for n large enough, with θ independent of n. Hence

$$(11.21) \qquad \int_{\Gamma_{n2}} \frac{|1 - t^2|^\eta \ E_{n\infty}(W)}{|t - |x|| \ |T_{n\infty}(W,a_nt)W(a_n|t|)|} \ |dt| \le C_7 \ \theta^n/(1 + |x|),$$

for n large enough, where C_7 is independent of n and x. To estimate the integral over Γ_{n1}, let $t := (\xi_n^2 + iy)^{1/2}$, $0 < y \le \epsilon$. We have

$$\begin{aligned}|1 - t^2|^\eta &= ((1 - \xi_n^2)^2 + y^2)^{\eta/2} \\ &\le C_8 \ (\rho_n^2 + y^2)^{\eta/2} = C_8 \ \rho_n^\eta \ (1 + y^2\rho_n^{-2})^{\eta/2},\end{aligned}$$

by (11.8) and Corollary 8.2. Further since $\mathrm{Re}(t) \ge 1/2$, n large enough,

$$|t - |x|| = |t^2 - x^2|/|t + |x||$$

$$\geq |\xi_n^2 + iy - x^2|/(1 + |x| + 2\epsilon)$$

$$\geq ((\xi_n^2 - x^2)^2 + y^2)^{1/2}/(2 + |x|)$$

$$\geq C_9(|\xi_n - |x|| + y),$$

with C_9 independent of n,y and x. Further,

$$\left|\frac{dt}{dy}\right| = \frac{1}{2}\,|\xi_n^2 + iy|^{-1/2} \leq 1, \quad 0 < y \leq \epsilon,$$

n large enough. Finally by (10.13), if ϵ is small enough,

$$E_{n\infty}(W) \,/\, |T_{n\infty}(W,a_n t)W(a_n|t|)| \leq (1 + C_{10}y^2\rho_n^{-2})^{-l},$$

where l is any given, but fixed, positive integer and C_{10} is independent of n,y and t. We assume now that $l \geq \eta/2 + 1$. Then

$$(11.22) \qquad I_{n1} := \int_{\Gamma_{n1}} \frac{|1 - t^2|^{\eta}\, E_{n\infty}(W)}{|t - |x||\; |T_{n\infty}(W,a_n t)W(a_n|t|)|}\, |dt|$$

$$\leq C_{11}\, \rho_n^{\eta} \int_0^{\epsilon} (|\xi_n - |x|| + y)^{-1}(1 + (y/\rho_n)^2)^{-l+\eta/2}\, dy$$

$$= C_{11}\, \rho_n^{\eta} \int_0^{\epsilon/\rho_n} (|\xi_n - |x||/\rho_n + u)^{-1}(1 + u^2)^{-l+\eta/2}\, du.$$

If $|\xi_n - |x||/\rho_n \geq 1$, we bound above the integral in this last right-hand side by

$$(|\xi_n - |x||/\rho_n)^{-1} \int_0^{\epsilon/\rho_n} (1 + u^2)^{-l+\eta/2}\, du$$

$$\leq C_{12}\rho_n/|\xi_n - x| \leq C_{13} \log\, (1 + \rho_n/|\xi_n - x|).$$

If $|\xi_n - |x||/\rho_n < 1$, we bound above the integral in the last right-hand side of (11.22) by

$$\int_0^1 (|\xi_n - |x||/\rho_n + u)^{-1}\, du + \int_1^{\infty} (1 + u^2)^{-l+\eta/2} du$$

$$\leq C_{14} \log\, (1 + \rho_n/|\xi_n - |x||) + C_{15}$$

$$\leq C_{16} \log\, (1 + \rho_n/|\xi_n - |x||).$$

Thus

$$I_{n1} \leq C_{17}\, \rho_n^{\eta} \log\, (1 + \rho_n/|\xi_n - |x||)$$

and so together with (11.21) and (11.16), this yields

$$I_n(x) \le C_{18}\, \rho_n^{\eta} \log (1 + \rho_n/|\xi_n - |x||).$$

Then Lemma 11.4 yields

(11.23) $|g(x) - L_n(x)/H(a_n x)| \le C_{19}\, \rho_n^{\eta}\, \|h\| \log (1 + \rho_n/|\xi_n - |x||)$, $|x| < \xi_n$ and

(11.24) $|L_n(x)/H(a_n x)| \le C_{20}\, \rho_n^{\eta}\, \|h\| \log(1 + \rho_n/|\xi_n - |x||)$, $|x| > \xi_n$.

These two inequalities yield (11.10), except for $|x| \in (\xi_n, 1)$. But for such x,

(11.25) $|g(x)| \le C_{21}\, \|h\|\, (1 - x^2)^{\eta} \le C_{22}\, \rho_n^{\eta}$,

by Corollary 8.2. Further, for such x,

$$\log (1 + \rho_n/|\xi_n - |x||) \ge \log (1 + \rho_n/(1 - \xi_n)) \ge C_{23},$$

by Corollary 8.2. Thus

$$|g(x)| \le C_{24}\, \|h\|\, \rho_n^{\eta} \log (1 + \rho_n/|\xi_n - |x||),\ |x| \in (\xi_n, 1).$$

This last inequality and (11.24) yield (11.10) for $|x| \in (\xi_n, 1)$. □

In the proof of (11.11), we shall need a crude Markov-Bernstein inequality. More precise inequalities have been obtained by Freud [14], with subsequent generalizations and filling of gaps by Levin and Lubinsky [22,23], and Nevai and Totik [58].

Lemma 11.5

Let $W(x) := e^{-Q(x)}$ be even and continuous in $\mathbb{R}$, and let $Q(x)$ be strictly increasing in $(0,\infty)$ with

$$\lim_{|x|\to\infty} \log Q(x) / \log |x| = \infty.$$

Then for $P \in \mathcal{P}_n$ and $n \ge 1$,

(11.26) $\|P'W\|_{L_\infty(\mathbb{R})} \le n^2 \{W(0)/W(2)\} \|PW\|_{L_\infty(\mathbb{R})}$.

Proof

Note the first classical Markov-Bernstein inequality [24,54]

$$\|P'\|_{L_\infty[-1,1]} \le n^2 \|P\|_{L_\infty[-1,1]},\ P \in \mathcal{P}_n.$$

Let $P \in \mathcal{P}_n$ and let ξ be such that

$$|P'W|(\xi) = \|P'W\|_{L_\infty(\mathbb{R})}.$$

Suppose first that $\xi \geq 2$. Then

$$\|P'W\|_{L_\infty(\mathbb{R})} = |P'W|(\xi)$$
$$\leq W(\xi)\, n^2\, \|P\|_{L_\infty[\xi-2,\xi]} \leq n^2 \|PW\|_{L_\infty[\xi-2,\xi]},$$

by monotonicity of W. If $0 \leq \xi \leq 2$, we obtain similarly

$$\|P'W\|_{L_\infty(\mathbb{R})} = |P'W|(\xi)$$
$$\leq n^2\, W(0)\, \|P\|_{L_\infty[0,2]} \leq n^2\{W(0)/W(2)\}\, \|PW\|_{L_\infty[0,2]}.$$

Similarly if $\xi < 0$. □

Proof of (11.11) in Theorem 11.2

Let $K > 0$. Note first that (11.10) implies an inequality like that in (11.11), if we restrict x so that $||x| - \xi_n| \geq n^{-K}$. We must estimate $g(x) - P_n(x)/H(a_nx)$ for the remaining x. Now, from (11.10), we see that

$$|P_n(x)/H(a_nx)| \leq C_7\, \|h\|\, (\rho_n^\eta \log n + 1),\ ||x| - \xi_n| \geq n^{-K},$$

where C_7 is independent of n, h and x. Then from (11.4),

(11.27) $\quad |P_n(x)W(a_nx)| \leq C_8\, \|h\|\, (\rho_n^\eta \log n + 1),\ ||x| - \xi_n| \geq n^{-K}.$

By choosing K large enough, we ensure that in the remaining interval of length $2n^{-K}$, $P_n(x)W(a_nx)$ cannot grow by more than $1+o(1)$, and so remains bounded by $C\, \|h\|\, (\rho_n^\eta \log n + 1)$ in $\mathbb{R}$. To prove this, we use (7.5) in Theorem 7.1 and (6.29) with $R := (\xi_n - n^{-K})a_n$. By (7.5), if $\epsilon := |u| - 1 > 0$, then

$$|P_n(uR/a_n)W(Ru)| \leq \|P_n(uR/a_n)W(Ru)\|_{L_\infty[-1,1]} \exp(n\, U_{n,R}(u))$$
$$\leq \|P_n(x)W(a_nx)\|_{L_\infty[-R/a_n, R/a_n]} \exp(-nC_9\epsilon^{3/2} + nC_{10}\epsilon^{1/2}(1 - R/a_n)).$$

Since Corollary 8.2 shows that

$$R/a_n \geq 1 - C_{11}((\log n)/n)^{2/3} - n^{-K} \geq 1 - C_{12}((\log n)/n)^{2/3},$$

and since (11.27) holds, we obtain if $\epsilon = O(n^{-K})$, some $K > 2/3$, that

$$|P_n(uR/a_n)W(Ru)| \leq 2C_8\, \|h\|\, \{\rho_n^\eta \log n + 1\}.$$

Setting $Ru = a_nx$ gives

$$|P_n(x)W(a_nx)| \le 2C_8 \,\|h\|\, \{\rho_n^\eta \log n + 1\},$$

for $R/a_n \le |x| \le (R/a_n)(1 + O(n^{-K}))$, that is

$$\xi_n - n^{-K} \le |x| \le (\xi_n - n^{-K})(1 + O(n^{-K})).$$

Choosing the $O(n^{-K})$ suitably covers the range

$$\xi_n - n^{-K} \le |x| \le \xi_n + n^{-K}.$$

Then by (11.27),

$$|P_n(x)W(a_nx)| \le 2\, C_8 \,\|h\|\, (\rho_n^\eta \log n + 1),\ x \in \mathbb{R}.$$

Using the Markov-Bernstein inequality Lemma 11.5, we can bound the derivative of $P_n(x)W(a_nx)$ by a power of n. Expanding $P_n(x)W(a_nx)$ about $x_n := \xi_n \pm n^{-K}$, we obtain if K is large enough,

$$|P_n(x)W(a_nx)| = |P_n(x_n)W(a_nx_n)| + O(\|h\|\, \rho_n^\eta \log n),$$

uniformly for $||x| - \xi_n| \le n^{-K}$. Also if K is large enough, we have as at (11.25),

$$|g(x)| \le C_{22} \,\|h\|\, \rho_n^\eta\ ,\ ||x| - \xi_n| \le n^{-K},$$

and so from (11.10) and (11.4),

$$|P_n(x_n)W(a_nx_n)| \le C_{15} \,\|h\|\, \rho_n^\eta \log n.$$

We deduce that for $||x| - \xi_n| \le n^{-K}$,

$$|g(x) - P_n(x)/H(a_nx)| \le C_{16} \,\|h\|\, \rho_n^\eta \log n. \ \square$$

Proof of Theorem 11.1

We first prove this in the case where all $k_n = 0$. Then Theorem 11.1 is a corollary of Theorem 11.2, since any function that is continuous in $\mathbb{R}$ and vanishes outside (-1,1) may be approximated by a function of the form

$$\hat{g}(x) = \begin{cases} (1 - x^2)\ h(x)\ , & x \in (-1,1), \\ 0 & , x \in \mathbb{R}\backslash(-1,1), \end{cases}$$

with h(x) a polynomial. The fact that we may choose $H = G_{Q/2}$ follows from Theorem 6 in [27,p.301], where it is shown that

$$G_Q(x) \sim \exp(2Q(x)),\ |x| \text{ large},$$

and so

$$G_{Q/2}(x) \sim \exp(Q(x)),\ |x| \text{ large}:$$

Our condition (11.3) on Q, guarantees that $W = e^{-Q}$ is a Freud weight in the sense of [27], satisfying (6) and (7) in [27,p.299].

To prove the result for general $\{k_n\}_1^\infty$, we shall use a device often used in [32] and that we shall use often in section 12. We have $P_n \in \mathcal{P}_n$ satisfying (11.6). Now replace n by $n-k_n$ in (11.6):

$$\lim_{n\to\infty} \|g(x) - P_{n-k_n}(x)/H(a_{n-k_n}x)\|_{L_\infty(\mathbb{R})} = 0.$$

Now make the substitutions $a_{n-k_n}x = a_n u$; $P_{n-k_n}(x) =: \hat{P}_n(u) \in \mathcal{P}_{n-k_n}$.

This yields

$$\lim_{n\to\infty} \|g(ua_n/a_{n-k_n}) - \hat{P}_n(u)/H(a_n u)\|_{L_\infty(\mathbb{R})} = 0.$$

By Lemma 8.5,

$$a_n/a_{n-k_n} = 1 + O(k_n/n) = 1 + o(1),$$

and the continuity of g in [-1,1] yields

$$\lim_{n\to\infty} \|g(u) - g(ua_n/a_{n-k_n})\|_{L_\infty(\mathbb{R})} = 0.$$

Then (11.6) follows in the general case. □

12. Approximation by Certain Weighted Polynomials, II.

Whereas in the previous section we considered approximation by weighted polynomials of the form $P_n(x)/H(a_n x)$, in this section we consider approximation by $P_n(x)W(a_n x)$, or more generally $P_n(x)W(c_n x)$. This requires replacing the reciprocal of the entire function $H(x)$ by the weight itself. In the case when $W(x) = \exp(-x^m)$, m a positive even integer, one can choose $H = 1/W$, but in the general case, we have to choose $H = G_{Q/2}$. Furthermore, a considerable amount of effort is involved in the transition from $1/H$ to W. First, an analogue of Theorem 11.1:

Theorem 12.1

Let $W(x) := e^{-Q(x)}$, *where* $Q(x)$ *is even and continuous in* $\mathbb{R}$, $Q''(x)$ *exists and is continuous in* $(0,\infty)$, *and* $Q'(x)$ *is positive in* $(0,\infty)$, *while for some* $C_1, C_2 > 0$,

$$(12.1) \qquad C_1 \le (xQ'(x))'/Q'(x) \le C_2, \qquad x \in (0,\infty).$$

Suppose, further, that $Q'''(x)$ *exists and is continuous for* x *large enough, with*

$$(12.2) \qquad x^2|Q'''(x)|/Q'(x) \le C_3, \qquad x \in (C_4,\infty).$$

Let $a_n = a_n(W)$ *for* $n = 1,2,3,\ldots$, *and let*

$$(12.3) \qquad c_n := a_n(1+\epsilon_n),\ n = 1,2,3,\ldots,$$

where $\{\epsilon_n\}_1^\infty$ *is a sequence of real numbers satisfying*

$$(12.4) \qquad \limsup_{n\to\infty} \epsilon_n n^{1/2} \le 0$$

and

$$(12.5) \qquad \liminf_{n\to\infty} \epsilon_n (\log n)^2 \ge 0.$$

Let $\{k_n\}_1^\infty$, $k_n \le n$, *be a sequence of non-negative integers satisfying*

$$(12.6) \qquad \lim_{n\to\infty} k_n n^{-1/2} = 0.$$

<u>Let</u> $g(x)$ <u>be positive and continuous in</u> $[-1,1]$. <u>Then there exists</u> $P_n \in \mathcal{P}_{n-k_n}$, $n=1,2,3,\ldots$, <u>satisfying</u>

$$(12.7) \qquad \|P_n(x)W(c_nx)\|_{L_\infty[-1,1]} \le C_5, \qquad n=1,2,3,\ldots,$$

<u>such that</u>

$$(12.8) \qquad \lim_{n\to\infty} P_n(x)W(c_nx) = g(x),$$

<u>uniformly in compact subsets of</u> $\{x: 0 < |x| < 1\}$, and

$$(12.9) \qquad \lim_{n\to\infty} P_n(x)W(c_nx) = 0,$$

<u>uniformly in closed subsets of</u> $\{x: |x| > 1\}$. <u>Furthermore</u>,

$$(12.10) \qquad \lim_{n\to\infty} \int_{-1}^{1} |\log |P_n(x)W(c_nx)/g(x)|| \, (1 - x^2)^{-1/2} dx = 0.$$

<u>Finally, if</u> $\{k_n\}_1^\infty$ <u>and</u> $\{\epsilon_n\}_1^\infty$ <u>are further restricted so that</u>

$$(12.11) \qquad \lim_{n\to\infty} \epsilon_n (n/(\log n))^{2/3} = -\infty,$$

<u>and</u>

$$(12.12) \qquad \limsup_{n\to\infty} k_n n^{-1/3} (\log n)^{-2/3} < \infty,$$

<u>we may also ensure that</u>

$$(12.13) \qquad |P_n(x)W(c_nx)| \ge C_6, \qquad x \in [-1,1], \; n=1,2,3,\ldots .$$

We remark that the polynomial above has complex coefficients. The reason for this is that we add a small imaginary part to ensure that $\log |P_n(x)W(c_nx)|$ is not too small near $x = \pm 1$. One may also think of $|P_n(x)|$ as $S_{2n}(x)^{1/2}$, where $S_{2n}(x)$ is a polynomial non-negative in $\mathbb{R}$, of degree at most $2(n - k_n)$. The following result establishes a certain precise interval in which we may approximate positive continuous functions by weighted polynomials $P_n(x)W(a_nx)$.

Theorem 12.2

Let $W(x)$ be as in Theorem 12.1, with the additional assumption that

(12.14) $\theta_n := a_n/n \to 0, \quad n \to \infty.$

Let $\{k_n\}_1^\infty$ be a sequence of non-negative integers satisfying (12.12), and let

(12.15) $\rho_n := ((\log n)/n)^{2/3}, \quad n = 2,3,4,\ldots .$

Let $g(x)$ be positive and continuous in $[-1,1]$ and let $\{r_n\}_1^\infty$ be a sequence of real numbers satisfying

(12.16) $\lim_{n\to\infty} r_n = \infty.$

Then there exists $P_n \in \mathcal{P}_{n-k_n}$, $n=1,2,3,\ldots$, such that

(12.17) $\lim_{n\to\infty} \|g(x) - P_n(x)W(a_nx)\|_{L_\infty(|x|\le 1-r_n(\theta_n+\rho_n))} = 0.$

Following is an analogue of Theorem 11.2.

Theorem 12.3

Let $W(x)$ be as in Theorem 12.1, and let $\{\theta_n\}_1^\infty$, $\{k_n\}_1^\infty$, and $\{\rho_n\}_1^\infty$ satisfy respectively (12.14), (12.12) and (12.15). Let $h(t)$ be analytic in $\{t \in \mathbb{C}: |t| \le 2\}$, let $\eta > 0$, and

(12.18) $$g(x) := \begin{cases} h(x)\,(1-x^2)^\eta, & x \in (-1,1), \\ 0, & x \in \mathbb{R}\backslash(-1,1). \end{cases}$$

Then there exists $P_n \in \mathcal{P}_{n-k_n}$, $n=1,2,3,\ldots$, such that

(12.19) $\|g(x) - P_n(x)W(a_nx)\|_{L_\infty(\mathbb{R})} \le C\{\rho_n^\eta \log n + (k_n/n)^{\hat{\eta}} + \theta_n^{\hat{\eta}\sigma/(\hat{\eta}+\sigma)}\},$

where

(12.20) $\hat{\eta} := \min\{1,\eta\}$; $\sigma := \min\{C_1, 1/4\}$,

and C_1 is the same constant as in (12.1). If, further, $h(x)$ is positive in $[-1,1]$, we may ensure that

$$(12.21)\qquad \lim_{n\to\infty} \int_{-1}^{1} |\log |P_n(x)W(a_nx)/g(x)|| \ (1-x^2)^{-1/2}\, dx = 0.$$

Note that Theorems 4.1 and 4.2 are immediate consequences of Theorems 12.1 and 12.3 respectively. Before giving the proofs of the latter results, we need a lemma on $G_{Q/2}$.

__Lemma 12.4__

__Let__ $W(x) := e^{-Q(x)}$ __be as in Theorem 12.1. Let__ $G_{Q/2}(x)$ __be the function defined by (11.2). Then__

$$(12.22)\qquad G_{Q/2}(x) = \{\pi\ T(x)\}^{1/2} W(x)^{-1}\ \{1 + O[Q(x)^{-1/2}\ (\log Q(x))^{3/2}]\},$$

__as__ $|x| \to \infty$, __where__

$$(12.23)\qquad T(x) := 1 + xQ''(x)/Q'(x),\ x \in \mathbb{R}\backslash\{0\}.$$

__Furthermore, if__

$$(12.24)\qquad \Psi(x) := 1/\{G_{Q/2}(x)W(x)\},\ x \in \mathbb{R},$$

__then__ $\Psi(x)$ __is continuous in__ $\mathbb{R}$, __and if__

$$(12.25)\qquad w_n(h;\delta) := \sup_{\substack{|x|,|y|\geq\delta \\ |x-y|\leq h}} |\Psi(a_nx) - \Psi(a_ny)|,$$

__then__

$$(12.26)\qquad w_n(h;\delta_n) \leq C \max\ \{h/\delta_n,\ (\log Q(a_n\delta_n))^{3/2} Q(a_n\delta_n)^{-1/2}\},$$

__uniformly for__ $h \geq 0$, $n \geq n_1$, __and any sequence__ $\{\delta_n\}_1^\infty$ __satisfying__

$$(12.27)\qquad a_n\delta_n \geq K.$$

__Here__ n_1, C __and__ K __are independent of__ n, h __and__ $\{\delta_n\}_1^\infty$.

__Proof__

The conditions on W(x) ensure that W(x)dx is a smooth Freud weight in the sense of [27]. More precisely, (6),(7) and (8) in [27,p.299] are satisfied with Q replaced by Q/2. Then by (21) in [27,p.301], we have that (12.22) is valid.

The continuity of $\Psi(x)$ is immediate; it is also clearly twice continuously differentiable in $(0,\infty)$. From (12.22), we may write

$$(12.28)\qquad \Psi(x) = \{\pi\ T(x)\}^{-1/2}\ Z(x),$$

where $Z(x)$ is continuous in $(0,\infty)$, bounded above and below there, and

$$(12.29)\qquad Z(x) = 1 + O\{\ (\log Q(x))^{3/2}\ Q(x)^{-1/2}\ \},\ |x| \to \infty.$$

Now

$$(12.30)\qquad \begin{aligned} T'(x) &= Q''(x)/Q'(x) + xQ'''(x)/Q'(x) - x(Q''(x)/Q'(x))^2 \\ &= O(1/x), \qquad |x| \to \infty, \end{aligned}$$

by (12.1) and (12.2), while we see that

$$(12.31)\qquad T(x) = (xQ'(x))'/Q'(x) \in (C_1,C_2),\ |x| \in (0,\infty).$$

Then there exists $K > 0$ such that if $a_n x \geq K$ and $h \geq 0$, then for some C independent of x and h, and $\xi \in (x,x + h)$,

$$(12.32)\qquad \begin{aligned} |T(a_n(x + h)) - T(a_n x)| &= a_n h\ |T'(a_n\xi)| \\ &\leq Ca_n h/(a_n x) = Ch/x. \end{aligned}$$

Next, if $a_n x \geq K$ for some large enough K, and $h \geq 0$, it follows from (12.29) that

$$(12.33)\qquad |Z(a_n(x + h)) - Z(a_n x)| \leq C^*\ (\log Q(a_n x))^{3/2}\ Q(a_n x)^{-1/2}.$$

Combining (12.28),(12.32) and (12.33), together with the fact that

$$\Psi(x) \sim T(x) \sim Z(x) \sim 1,\ x \in (0,\infty),$$

we obtain (12.26). □

Next, we use Lemma 6.3 to construct weighted polynomials whose logarithm is not too small. Here the assumptions (12.4) and (12.5) are crucial.

<u>Lemma 12.5</u>

<u>Let</u> $W,\{c_n\},\{k_n\}$ <u>and</u> $\{\epsilon_n\}$ <u>be as in Theorem 12.1 (in particular, the</u>

ϵ_n's <u>satisfy</u> (12.4) <u>and</u> (12.5)). <u>Then there exist</u> $R_n \in P_{n-k_n}$, $n=1,2,3,\ldots$, <u>such that</u>

(12.34) $$\|R_n(x)W(c_nx)\|_{L_\infty(\mathbb{R})} \le Cn^{-2},\ n=1,2,3,\ldots,$$

<u>and if</u> $\{A_n\}_1^\infty$ <u>is any sequence of subsets of</u> $[-1,1]$ <u>satisfying</u>

(12.35) $$\operatorname{meas}(A_n) = o((\log n)^{-2}),\ n \to \infty,$$

<u>then</u>

(12.36) $$\lim_{n\to\infty} \int_{A_n} |\log |R_n(x)W(c_nx)|| (1 - x^2)^{-1/2}\, dx = 0.$$

<u>Proof</u>

For $n=1,2,3,\ldots$, let $l_n := n-k_n$, $b_n := c_n/a_{l_n}$, and $\hat{b}_n := \min\{1, b_n^{-1}\}$. Note that by (12.3), (12.4), (12.6) and Lemma 8.5,

(12.37) $$\begin{aligned} b_n &= (a_n/a_{l_n})(1 + \epsilon_n) \\ &\le (1 + O(k_n/n))(1 + \epsilon_n) \\ &\le 1 + \eta_n n^{-1/2}, \end{aligned}$$

where

$$\lim_{n\to\infty} \eta_n = 0.$$

Now let $\hat{P}_{l_n}(x)$ denote the polynomial of Lemma 6.3, with $R = a_{l_n}$, having simple zeros $y_1, y_2, \ldots, y_{l_n}$, say, in $(-1,1)$. Let C_5 be as in Lemma 6.3 and

$$R_n(x) := \hat{P}_{l_n}(b_nx)\, n^{-C_5-2},\ n=1,2,3,\ldots .$$

Then, by the substitution $c_nx = a_{l_n}u$,

$$\begin{aligned} \|R_n(x)W(c_nx)\|_{L_\infty(\mathbb{R})} &= \|\hat{P}_{l_n}(u)W(a_{l_n}u)\|_{L_\infty(\mathbb{R})}\, n^{-C_5-2} \\ &= \|\hat{P}_{l_n}(u)W(a_{l_n}u)\|_{L_\infty[-1,1]}\, n^{-C_5-2} = O(n^{-2}), \end{aligned}$$

by (6.32). Thus (12.34) is valid. Next, let

$$F_{jn} := \{x : |b_nx - y_j| \le n^{-8}\},\ j = 1,2,\ldots,l_n,$$

and

$$F_n := \left(\bigcup_{j=1}^{l_n} F_{jn} \right) \cap [-\hat{b}_n, \hat{b}_n].$$

We have for $x \in G_n := [-\hat{b}_n, \hat{b}_n] \backslash F_n$, from (6.32),

$$| \log |R_n(x)W(c_nx)|| \le C_7 \log n,$$

and so

$$(12.38) \quad \int_{A_n \cap G_n} |\log |R_n(x)W(c_nx)|| \, (1 - x^2)^{-1/2} \, dx$$
$$\le C_8(\log n) \, \{\mathrm{meas}(A_n \cap G_n)\}^{1/2} \to 0, \ n \to \infty,$$

by (12.35). Next, if $x \in F_{jn}$ and y_j is the closest among $\{y_1, y_2, \ldots, y_{l_n}\}$ to b_nx, while $|x| \le \hat{b}_n$,

$$|\log |R_n(x)W(c_nx)|| = |\log |\hat{P}_{l_n}(b_nx) \, W(a_{l_n} b_nx) \, n^{-C_5-2}||$$
$$\le C_9 \, |\log |b_nx - y_j||.$$

Here,

$$\int_{F_{jn}} |\log |b_nx - y_j|| \, (1 - x^2)^{-1/2} \, dx$$
$$\le C_{10} \max \left\{ \int_{|bx-y| \le n^{-8}} |\log|bx - y|| \, (1 - x^2)^{-1/2} \, dx: \ b \in [1/2, 2], \ y \in [-1, 1] \right\}.$$
$$\le C_{11} \int_0^{n^{-8}} |\log u| \, u^{-1/2} du \le C_{12} \, n^{-3}.$$

It follows that

$$\int_{F_n} |\log |R_n(x)W(c_nx)|| \, (1 - x^2)^{-1/2} \, dx \le C_{12} \, n^{-2}.$$

Together with the definition of F_n, G_n and (12.38), this last inequality shows that

$$(12.39) \quad \lim_{n \to \infty} \int_{A_n \cap [-\hat{b}_n, \hat{b}_n]} |\log|R_n(x)W(c_nx)|| \, (1 - x^2)^{-1/2} \, dx = 0.$$

This establishes the result if all $\hat{b}_n \ge 1$. Suppose now for infinitely many n, $\hat{b}_n \le 1$; that is, $b_n = c_n/a_{l_n} \ge 1$. For $\hat{b}_n = b_n^{-1} \le |x| \le 1$, we have from (6.33) and (6.29), that

$$\begin{aligned}\log |R_n(x)W(c_nx)| &= \log |\hat{P}_{l_n}(b_nx)W(a_{l_n}b_nx)| + O(\log n),\\ &= nU_{n,a_{l_n}}(b_nx) + O(\log n)\\ &\geq -n\,C_{13}\,(b_nx-1)^{3/2} + O(\log n)\\ &\geq -C_{14}\,[n^{1/4}\,\eta_n^{3/2} + \log n],\end{aligned}$$

by (12.37). A similar upper bound follows from (6.29). Then

$$\int_{\hat{b}_n\leq|x|\leq 1} |\log |R_n(x)W(c_nx)||\,(1-x^2)^{-1/2}\,dx$$

$$\leq C_{14}\,(n^{1/4}\,\eta_n^{3/2} + \log n)\int_{b_n^{-1}}^{1}(1-x^2)^{-1/2}\,dx \to 0,\ n\to\infty,$$

by (12.37) again. Together with (12.39), this yields (12.36). □

Proof of (12.7) to (12.10) of Theorem 12.1

This consists of three steps: First we approximate by weighted polynomials of the form $P_n(u)/G_{Q/2}(a_nu)$. Then we approximate $\Psi(a_nu) = 1/\{G_Q(a_nu)W(a_nu)\}$ by a low degree polynomial. Finally, we add a small imaginary part to ensure that the log of the weighted polynomial is not too small near ± 1, ensuring (12.10). We shall deal with (12.11), (12.12) separately.

Let $h(x)$ be a polynomial positive in $[-1,1]$ and let

$$(12.40)\qquad \hat{g}(x) := \begin{cases} h(x)\,(1-x^2) & ,x\in(-1,1),\\ 0 & ,x\in\mathbb{R}\setminus(-1,1).\end{cases}$$

We first find polynomials satisfying (12.7) to (12.10) when g is replaced by $\hat{g}$.

Step 1

By Theorem 11.2, with $H = G_{Q/2}$, we can find polynomials $P_n^*(u)$ of degree at most n, such that

$$\|\hat{g}(u) - P_n^*(u)/G_{Q/2}(a_nu)\|_{L_\infty(\mathbb{R})} \leq C_6\,\rho_n\,\log n,$$

where ρ_n is given by (11.8). Now let l be a positive integer satisfying $1 \leq l \leq n/2$. Making the substitution $a_n x = a_{n-l} u$ in the inequality

$$\|\hat{g}(u) - P^*_{n-l}(u)/G_{Q/2}(a_{n-l}u)\|_{L_\infty(\mathbb{R})} \leq C_6 \rho_{n-l} \log (n - l),$$

we obtain for the polynomial $\hat{P}_{n-l}(x) := P^*_{n-l}(u)$,

$$\|\hat{g}(a_n x/a_{n-l}) - \hat{P}_{n-l}(x)/G_{Q/2}(a_n x)\|_{L_\infty(\mathbb{R})} \leq C_7 \rho_n \log n.$$

Here, as $\hat{g}(x)$ is continuously differentiable in $(-1,1)$,

$$\|\hat{g}(a_n x/a_{n-l}) - \hat{g}(x)\|_{L_\infty(\mathbb{R})} \leq C_8 \, |a_n/a_{n-l} - 1| \leq C_9 \, l/n,$$

by (8.13) in Lemma 8.5. Thus

$$\begin{aligned}\|\hat{g}(x) - \hat{P}_{n-l}(x)/G_{Q/2}(a_n x)\|_{L_\infty(\mathbb{R})} &\leq C_{10}(\rho_n \log n + l/n) \\ &\leq C_{11} \rho_n \log n,\end{aligned}$$

if

(12.41) $\quad l \leq n^{1/3}.$

Using (12.24), we have

(12.42) $\quad \|\hat{g}(x) - \hat{P}_{n-l}(x) W(a_n x) \Psi(a_n x)\|_{L_\infty(\mathbb{R})} \leq C_{11} \rho_n \log n.$

Step 2

Next, we approximate $\Psi(a_n x)$ by a low degree polynomial. Unfortunately, we are unable to do this near $x = 0$, at least in general; so we define

(12.43) $$\hat{\Psi}(a_n x) := \begin{cases} \Psi(a_n x), & |x| \geq 1/\log n, \\ \Psi(a_n/\log n), & |x| \leq 1/\log n. \end{cases}$$

Then by (12.25) and (12.26) for $s \in \mathbb{R}$,

(12.44) $\quad |\hat{\Psi}(a_n(s + n^{-1/3})) - \hat{\Psi}(a_n s)| \leq w(n^{-1/3};\ 1/\log n) \to 0,\ n \to \infty,$

since Lemma 8.5 shows that

$$a_n/\log n \to \infty,\ n \to \infty.$$

Then, by Jackson's theorem, we can find polynomials $S_l(x)$ of degree $l \sim n^{1/3}$, such that

$$\|S_l(x) - \hat{\Psi}(a_n x)\|_{L_\infty[-2,2]} \to 0, \; n \to \infty.$$

Since $\Psi(x) \sim 1$ uniformly for $x \in \mathbb{R}$, the same is true of $\hat{\Psi}(x)$. Thus

$$(12.45) \qquad \|S_l(x)\hat{\Psi}^{-1}(a_n x) - 1\|_{L_\infty[-2,2]} \to 0, \; n \to \infty.$$

Letting

$$(12.46) \qquad X_n(x) := \hat{P}_{n-l}(x) S_l(x),$$

a polynomial of degree at most n, we have from (12.42)

$$\|\hat{g}(x) - X_n(x) W(a_n x) \Psi(a_n x) \hat{\Psi}^{-1}(a_n x)\|_{L_\infty[-2,2]} \to 0, \; n \to \infty.$$

Then, taking account of (12.43), we have

$$(12.47) \qquad \lim_{n\to\infty} \|\hat{g}(x) - X_n(x) W(a_n x)\|_{L_\infty(1/\log n \leq |x| \leq 2)} = 0.$$

Furthermore as Ψ and $\hat{\Psi} \sim 1$ in $\mathbb{R}$, for $n \geq n_1$,

$$\|X_n(x) W(a_n x)\|_{L_\infty[-2,2]} \leq C \|\hat{g}\|_{L_\infty[-2,2]},$$

and so for $n \geq n_1$,

$$(12.48) \qquad \|X_n(x) W(a_n x)\|_{L_\infty(\mathbb{R})} \leq C \|\hat{g}\|_{L_\infty(\mathbb{R})}.$$

Next, we establish a lower bound for $X_n(x)W(a_n x)$. First note from (12.45) that for $n \geq n_1$,

$$C_{12} \leq |S_l(x)| \leq C_{13}, \; x \in [-2,2].$$

Further, noting the form (12.40) of $\hat{g}$ and the positivity of h in $[-1,1]$, we have from (12.42),

$$|\hat{P}_{n-l}(x) W(a_n x)| \geq C_{14} \, (\hat{g}(x) - C_{11} \, \rho_n \log n)$$
$$\geq C_{15} \hat{g}(x), \qquad |x| \leq 1 - C_{16} \, \rho_n \log n.$$

Then for $n \geq n_1$,

$$(12.49) \qquad |X_n(x) W(a_n x)| \geq C_{17} \, \hat{g}(x), \qquad |x| \leq 1 - C_{16} \, \rho_n \log n.$$

Now let us make the substitutions

$$a_{n-k_n} x = c_n u; \quad \hat{X}_n(u) := X_{n-k_n}(x) \in \mathcal{P}_{n-k_n}.$$

in (12.47), (12.48) and (12.49), after first replacing n by $n - k_n$ in those relations. Setting $b_n := c_n/a_{n-k_n}$, we obtain

$$\lim_{n\to\infty} \|\hat{g}(b_n u) - \hat{X}_n(u)W(c_n u)\|_{L_\infty(2/\log n \le |u| \le 3/2)} = 0,$$

(12.50) $$\|\hat{X}_n(u)W(c_n u)\|_{L_\infty(\mathbb{R})} \le C\|\hat{g}\|_{L_\infty(\mathbb{R})},$$

and

$$|\hat{X}_n(u)W(c_n u)| \ge C_{17}\, \hat{g}(b_n u), \qquad |u| \le b_n^{-1}(1 - C_{16}\, \rho_n \log n).$$

Now as in (12.37), in the proof of Lemma 12.4,

$$b_n \le 1 + o(n^{-1/2}).$$

Also, from (12.5),

$$b_n = (a_n/a_{n-k_n})(1 + \epsilon_n) \ge 1 + \epsilon_n \ge 1 - o((\log n)^{-2}).$$

Taking account of the form (12.40) of $\hat{g}(x)$, we obtain

(12.51) $$|\hat{X}_n(u)W(c_n u)| \ge C_{18}\, \hat{g}(u), \ |u| \le 1 - \delta_n(\log n)^{-2},$$

where

(12.52) $$\lim_{n\to\infty} \delta_n = 0,$$

and

(12.53) $$\lim_{n\to\infty} \|\hat{g}(u) - \hat{X}_n(u)W(c_n u)\|_{L_\infty(2/\log n \le |u| \le 3/2)} = 0.$$

<u>Step 3</u>

Let $R_n(u)$ be the polynomial of Lemma 12.5 and

$$P_n(u) := \hat{X}_n(u) + iR_n(u), \ n \text{ large enough}.$$

From (12.34) and (12.50), for $n \ge n_1$,

(12.54) $$\|P_n(u)W(c_n u)\|_{L_\infty(\mathbb{R})} \le 2C\, \|\hat{g}\|_{L_\infty(\mathbb{R})},$$

and from (12.34) and (12.53),

(12.55) $$\lim_{n\to\infty} \|\hat{g}(u) - P_n(u)W(c_n u)\|_{L_\infty(2/\log n \le |u| \le 3/2)} = 0,$$

while from (12.51),

$$|P_n(u)W(c_n u)| \ge C_{18}\, \hat{g}(u), \ |u| \le 1 - \delta_n\, (\log n)^{-2}.$$

Then, we deduce that

$$\lim_{n\to\infty} \int_{|u|\le 1-\delta_n(\log n)^{-2}} |\log|P_n(u)W(c_nu)/\hat{g}(u)||(1 - u^2)^{-1/2}du = 0.$$

Also, from Lemma 12.5,

$$\lim_{n\to\infty} \int_{1\ge|u|\ge 1-\delta_n(\log n)^{-2}} |\log|P_n(u)W(c_nu)/\hat{g}(u)||(1 - u^2)^{-1/2}du = 0.$$

It follows that we have established (12.7),(12.8) and (12.10) for $\hat{g}$, while (12.54),(12.3) and Lemma 7.4 yield (12.9). Since any g positive and continuous in [-1,1] can be approximated in a suitable sense by $\hat{g}$ of the form (12.40), the result follows. □

Proof of (12.13) when (12.11) and (12.12) hold.

We have to modify the previous proof. It suffices to consider the case when g(x) is a polynomial positive in [-1,1]. Outside [-1,1], we set g(x) := 0. Note that (12.13),(12.7) and (12.8) immediately yield (12.10). We proceed in two steps.

Step 1

By Theorem 11.2, with $H := G_{Q/2}$, $\eta = 0$ and ρ_n and ξ_n as in (11.8) and (11.9) respectively, there exist polynomials $P_n^*(u)$ of degree at most n such that

$$|g(u) - P_n^*(u)/G_{Q/2}(a_nu)| \le C_5 \log(1 + \rho_n/||u| - \xi_n|),$$

$u \in \mathbb{R}$, n=1,2,3,... . By Corollary 8.2,

$$1 \ge \xi_n \ge 1 - C_6\rho_n.$$

Then if

$$\lim_{n\to\infty} r_n = \infty, \tag{12.56}$$

we have

$$|g(u) - P_n^*(u)/G_{Q/2}(a_nu)| \le C_8/r_n, \quad |u| \le 1- C_9r_n\rho_n. \tag{12.57}$$

Further, since g(u) is positive and continuous in [-1,1],

$$|P_n^*(u)/G_{Q/2}(a_nu)| \le C_{10}, \quad ||u| - 1| \ge C_{11}\rho_n. \tag{12.58}$$

and

$$(12.59) \qquad |P_n^*(u)/G_{Q/2}(a_n u)| \geq C_{11}, \quad |u| \leq 1 - C_{12}\rho_n.$$

To bound $P_n^*(u)/G_{Q/2}(a_n u)$ in the remaining intervals $||u| - 1| \leq C_{11}\rho_n$ in which (12.58) does not apply, we use (11.11), which shows that

$$(12.60) \qquad \|P_n^*(u)/G_{Q/2}(a_n u)\|_{L_\infty(\mathbb{R})} \leq C_{12} \log n.$$

Replacing n by $n-k_n-l$ in (12.57) to (12.60) and then making the substitutions

$$a_{n-k_n-l}u = a_n x; \qquad P^*_{n-k_n-l}(u) =: \hat{P}_n(x) \in \mathcal{P}_{n-k_n-l},$$

we obtain much as in Step 1 of the previous proof, that if $l \sim n^{1/3}$,

$$(12.61) \qquad |g(x) - \hat{P}_n(x)/G_{Q/2}(a_n x)| \leq C_{13}(r_n^{-1} + \rho_n), \quad |x| \leq 1 - C_{14}r_n\rho_n,$$

$$(12.62) \qquad C_{14} \geq |\hat{P}_n(x)/G_{Q/2}(a_n x)| \geq C_{15}, \quad |x| \leq 1 - C_{16}\rho_n,$$

and

$$(12.63) \qquad \|\hat{P}_n(u)/G_{Q/2}(a_n u)\|_{L_\infty(\mathbb{R})} \leq C_{12} \log n,$$

noting that by (12.12),

$$(k_n + l)/n = O(\rho_n) + O(n^{-2/3}) = O(\rho_n).$$

<u>Step 2</u>

Let $S_l(x)$ denote the polynomial of degree $l \sim n^{1/3}$ of Step 2 of the previous proof satisfying (12.45), where $\hat{\Psi}$ is given by (12.43). Letting

$$P_n^\#(x) := \hat{P}_n(x)S_l(x) \in \mathcal{P}_{n-k_n},$$

we obtain from (12.43),(12.45) and (12.61),

$$\lim_{n\to\infty} \|g(x) - P_n^\#(x)W(a_n x)\|_{L_\infty(1/\log n \leq |x| \leq 1-C_{20}r_n\rho_n)} = 0,$$

while from (12.45) and (12.62),

$$C_{22} \geq |P_n^\#(x)W(a_n x)| \geq C_{21}, \quad |x| \leq 1 - C_{18}\rho_n,$$

and from (12.63),

(12.64) $\|P_n^{\#}(x)W(a_nx)\|_{L_\infty(\mathbb{R})} = \|P_n^{\#}(x)W(a_nx)\|_{L_\infty[-2,2]} \leq C_{12}\log n.$

If $c_n = a_n(1+\epsilon_n)$, where $\epsilon_n \leq -C_{18}\rho_n$, the substitutions $a_nx = c_nu$, $P_n(u) := P_n^{\#}(x)$, yield (12.7), (12.8) and (12.13), provided $r_n \to \infty$ sufficiently slowly.

To prove (12.9), we use (12.64) and (7.5), with $R = a_n$, and also the fact that for each $\delta > 0$,

$$\max\{\exp(nU_{n,a_n}(x)): |x| \geq 1+\delta\} \leq e^{-C(\delta)n},\ n \to \infty. \ \square$$

In the proof of Theorem 12.2, we shall use the following estimate on the modulus of continuity of Ψ:

Lemma 12.6

Let $W(x)$ *be as in Theorem 12.1, and let* $G_{Q/2}(x)$ *and* $\Psi(x)$ *be given by* (11.2) *and* (12.24) *respectively. Then*

(12.65) $G'_{Q/2}(x)/G_{Q/2}(x) \leq C_3(Q'(x)+1),\ x \in (0,\infty).$

Further, if

(12.66) $$w_n(h) := \sup_{\substack{0<s\leq h\\ s\in\mathbb{R}}} |\Psi(a_nx) - \Psi(a_n(x+s))|,$$

and

(12.67) $$\lim_{n\to\infty} a_n/n = 0,$$

and if C_1 *is as in* (12.1), *then*

(12.68) $$w_n(h) \leq C_4\max\{(a_nh)^{C_1}, Q(a_n\delta_n)h/\delta_n, (\log Q(a_n\delta_n))^{3/2}Q(a_n\delta_n)^{-1/2}\},$$

uniformly for $h \geq 0$, n *large enough, and any sequence* $\{\delta_n\}_1^\infty$ *satisfying*

(12.69) $a_n\delta_n \geq K.$

Here C_4, C_5 *and* K *are independent of* n, h *and* $\{\delta_n\}_1^\infty$.

Proof

Note first from (6.20) that $Q(x) \sim xQ'(x)$, x large enough. Hence for some C and $j \geq 1$,

(12.70) $\quad Q(q_j) \leq C\, q_j Q'(q_j) = Cj.$

Next, if $L \geq 2$, we have for $j \geq L$,

$$L = q_j Q'(q_j)/(q_{j/L} Q'(q_{j/L})).$$

By (6.19) with $x = q_{j/L}$, $t = q_j/q_{j/L}$, we have

$$L \leq (q_j/q_{j/L})^{C_2},$$

so

(12.71) $\quad q_{j/L}/q_j \leq L^{-1/C_2}.$

We now use (12.70) and (12.71) to estimate $G'_{Q/2}(x)$, much as in Levin and Lubinsky [23]. Now, by (11.2),

$$G'_{Q/2}(x) = \sum_{j=1}^{\infty} (x/q_{2j})^{2j}\, j^{-1/2}\, e^{Q(q_{2j})}\, 2j/x.$$

We let L be some large positive number, and split the sum on the last right-hand side into a sum $\sum_1$, taken over indices j for which $j \leq LxQ'(x)$, and a sum $\sum_2$ for indices j satisfying $j > LxQ'(x)$. We see that

$$\sum\nolimits_1 \leq 2LQ'(x)\, G_{Q/2}(x).$$

Now, if $j > LxQ'(x)$, then

$$q_{j/L} Q'(q_{j/L}) = j/L > xQ'(x),$$

and so $q_{j/L} > x$, by strict monotonicity of $tQ'(t)$. Then

$$\sum\nolimits_2 \leq \sum_{j > LxQ'(x)} (q_{j/L}/q_{2j})^{2j-1}\, j^{-1/2}\, e^{Q(q_{2j})}\, 2j$$

$$\leq 2 \sum_{j>0} (2L)^{-(2j-1)/C_2}\, j^{1/2}\, e^{C2j},$$

by (12.70) and (12.71). Since C and C_2 are independent of L,x, we may choose L independent of x and so large that

$$\sum\nolimits_2 \leq 1.$$

Thus

(12.72) $G'_{Q/2}(x) \le 2LQ'(x)G_{Q/2}(x) + 1,\ x > 0.$

Then (12.65) follows.

Now let us assume (12.67). Then (6.21) shows that

$$\lim_{n\to\infty} Q'(a_n) = \infty.$$

We shall use this to show that

(12.73) $$\lim_{x\to\infty} Q'(x) = \infty.$$

Indeed, it follows from (8.13) that $a_{2n} \sim a_n$, and from the monotonicity of $uQ'(u)$,

$$a_n \le x \le a_{2n} \text{ implies } Q'(x) \ge a_nQ'(a_n)/x \ge a_nQ'(a_n)/a_{2n} \ge C\, Q'(a_n).$$

Thus (12.73) follows.

Next, the definition (12.24) of Ψ, the fact that $\Psi \sim 1$ in $\mathbb{R}$, and (12.65), show that

$$|\Psi'(x)| \le C_7(Q'(x) + 1),\ x \in (0,\infty).$$

Hence if $0 < s \le h$ and $1 \le a_nx \le a_n(x + s) \le a_n\delta_n$,

(12.74) $$\begin{aligned}|\Psi(a_nx) - \Psi(a_n(x + s))| &\le C_8\ a_nh \max\{Q'(u): u \in [1,a_n\delta_n]\}\\ &\le C_9\ a_nh\ Q'(a_n\delta_n),\end{aligned}$$

if (12.69) holds with K large enough (by (12.73)). Also, (6.17) in Lemma 6.2 shows that Q is of Lipschitz class at least C_1 in [0,1]; so the same is true of Ψ. Then, for $0 < s \le h$ and $0 < a_nx \le a_n(x+s) \le 1$,

$$|\Psi(a_nx) - \Psi(a_n(x + s))| \le C_{10}(a_nh)^{C_1}.$$

Together with (12.25) and (12.26), this last inequality and (12.74) show that

$$w_n(h) \le C_{11}\max\{(a_nh)^{C_1},\ a_nhQ'(a_n\delta_n),\ h/\delta_n,(\log Q(a_n\delta_n))^{3/2}Q(a_n\delta_n)^{-1/2}\}.$$

Here, by (6.20), if K is large enough and $a_n\delta_n \ge K$,

$$a_nhQ'(a_n\delta_n) \sim Q(a_n\delta_n)h/\delta_n \ge C_{12}h/\delta_n.$$

Hence (12.68). □

We remark (see [20]) that one can show that

$$G'_{Q/2}(x)/G_{Q/2}(x) = Q'(x)\ \{1 + o(1)\},\ |x| \to \infty.$$

Proof of Theorem 12.2

It suffices to consider the case where g(x) is a polynomial that is positive in [-1,1]. We proceed much as in the proof of (12.13) of Theorem 12.1.

If $\{r_n\}_1^\infty$ satisfies (12.16), Theorem 11.2 yields polynomials $P_n^*(u)$ of degree at most n satisfying

$$|g(u) - P_n^*(u)/G_{Q/2}(a_n u)| \le C_8/r_n,\ |u| \le 1 - C_9 r_n \rho_n,$$

see (12.57). Replacing n by $n - k_n - l$ in this last inequality, and then making the substitutions

$$a_{n-k_n-l}u = a_n x\ ; \quad P^*_{n-k_n-l}(u) =: \hat{P}_n(x) \in \mathcal{P}_{n-k_n-l},$$

we obtain as at (12.61),

$$(12.75)\quad |g(x) - \hat{P}_n(x)/G_{Q/2}(a_n x)| \le C_{10}(r_n^{-1} + \rho_n + l/n),$$
$$|x| \le 1 - C_{11}(r_n \rho_n + l/n),$$

provided that $l = o(n)$, and since $k_n/n = O(\rho_n)$. We shall choose

$$(12.76)\quad l := l_n \sim r_n a_n,\ n \to \infty,$$

which we may do, if we assume that r_n grows so slowly that

$$(12.77)\quad r_n a_n = o(n),\ n \to \infty.$$

This we may assume without loss of generality. Then

$$(12.78)\quad \lim_{n\to\infty} \|g(x) - \hat{P}_n(x)/G_{Q/2}(a_n x)\|_{L_\infty(|x| \le 1 - C_{12} r_n(\rho_n + \theta_n))} = 0.$$

Next, we approximate $\Psi(a_n x)$ by a low degree polynomial, using the modulus of continuity estimates of Lemma 12.6. Choose an increasing sequence $\{\chi_n\}_1^\infty$ of positive numbers, satisfying

$$\lim_{n\to\infty} \chi_n = \infty,$$

but

$$\lim_{n\to\infty} Q(\chi_n)/(r_n\chi_n) = 0.$$

Let

$$\delta_n := \chi_n/a_n,\ n=1,2,3,\dots .$$

Then

$$(\log Q(a_n\delta_n))^{3/2}Q(a_n\delta_n)^{-1/2} \to 0,\ n \to \infty,$$

and if $h := 1/l_n$, (12.76) shows that

$$Q(a_n\delta_n)h/\delta_n = Q(\chi_n)a_n/(\chi_n l_n) \sim (Q(\chi_n)/\chi_n)\ r_n^{-1} \to 0,\ n \to \infty.$$

Finally,

$$a_n h = a_n/l_n \sim r_n^{-1} \to 0,\ n \to \infty.$$

Then, from (12.68),

$$w_n(1/l_n) \to 0,\ n \to \infty.$$

By Jackson's theorem, we can find a sequence of polynomials $S_{l_n}(x)$ of degree at most l_n satisfying

$$\lim_{n\to\infty} \|\Psi(a_nx) - S_{l_n}(x)\|_{L_\infty[-2,2]} = 0,$$

and so

(12.79) $$\lim_{n\to\infty} \|\Psi^{-1}(a_nx)S_{l_n}(x) - 1\|_{L_\infty[-2,2]} = 0.$$

Let

$$P_n(x) := \hat{P}_n(x)S_{l_n}(x),$$

a polynomial of degree at most $n - k_n$. From (12.78) and (12.79),

$$\lim_{n\to\infty} \|g(x) - P_n(x)W(a_nx)\|_{L_\infty(|x|\le 1-C_{13}r_n(\rho_n+\theta_n))}$$

$$= \lim_{n\to\infty} \|g(x) - \{\hat{P}_n(x)/G_{Q/2}(a_nx)\} \times \{S_{l_n}(x)\Psi^{-1}(a_nx)\}\|_{L_\infty(|x|\le 1-C_{13}r_n(\rho_n+\theta_n))} = 0.$$

Hence (12.17). □

<u>Proof of Theorem 12.3</u>

This is rather similar to the three previous proofs, and so we merely outline those steps that have appeared before. Firstly, Theorem 11.2

with $H = G_{Q/2}$ yields polynomials $P_n^*(u)$ of degree at most n, such that

$$\|g(u) - P_n^*(u)/G_{Q/2}(a_n u)\|_{L_\infty(\mathbb{R})} \le C_6\ \rho_n^\eta \log n.$$

Replacing n by $n - k_n - l$ in this last inequality, and then making the substitutions

$$a_{n-k_n-l}u = a_n x\ ;\ P_{n-k_n-l}^*(u) =: \hat{P}_n(x),$$

we obtain

$$(12.80)\qquad \|g(x) - \hat{P}_n(x)/G_{Q/2}(a_n x)\|_{L_\infty(\mathbb{R})} \le C_7(\rho_n^\eta \log n + (k_n/n)^{\hat{\eta}} + (l/n)^{\hat{\eta}}),$$

where $\hat{\eta} := \min\{1, \eta\}$. Let us now set

$$\sigma := \min\{C_1, 1/4\}\ ;\ \tau := \hat{\eta} / (\hat{\eta} + \sigma),$$

and choose $\{r_n\}_1^\infty$ and $\{l_n\}_1^\infty$ with (recall $\theta_n = a_n/n$)

$$(12.81)\qquad r_n \sim \theta_n^{-\tau},$$

$$(12.82)\qquad l_n \sim a_n r_n.$$

Further choose $\{\delta_n\}_1^\infty$ with

$$Q(a_n\delta_n) \sim \theta_n^{-3\tau/4}.$$

Setting $h := 1/l_n$ in (12.68), we have for n large enough,

$$\begin{aligned} w_n(1/l_n) &\le C_4 \max\{ (a_n/l_n)^{C_1},\ Q(a_n\delta_n)/(\delta_n l_n),\ Q(a_n\delta_n)^{-1/3} \} \\ &\le C_8 \max\{ \theta_n^{\tau C_1},\ \theta_n^{-3\tau/4}/(\delta_n a_n \theta_n^{-\tau}),\ \theta_n^{\tau/4} \} \\ &\le C_9\ \theta_n^{\tau\sigma}, \end{aligned}$$

since $a_n\delta_n \to \infty$, $n \to \infty$. By Jackson's theorem and Lemma 12.6 and the fact that $\Psi(x) \sim 1$ in $\mathbb{R}$, we can find polynomials $S_{l_n}(x)$ of degree l_n such that

$$(12.83)\qquad \|S_{l_n}(x)\Psi^{-1}(a_n x) - 1\|_{L_\infty[-2,2]} \le C_{12}\ \theta_n^{\tau\sigma}.$$

Now let

$$P_n(x) := \hat{P}_n(x) S_{l_n}(x),$$

a polynomial of degree at most $n - k_n$. From (12.80) and (12.83), we have

$$\|g(x) - P_n(x)W(a_nx)\|_{L_\infty[-2,2]}$$
$$\leq C_{13}\left(\rho_n^{\eta} \log n + (k_n/n)^{\hat{\eta}} + (a_nr_n/n)^{\hat{\eta}} + \theta_n^{\tau\sigma} \right)$$
$$\leq C_{14}\left(\rho_n^{\eta} \log n + (k_n/n)^{\hat{\eta}} + \theta_n^{(1-\tau)\hat{\eta}} + \theta_n^{\tau\sigma} \right),$$

by (12.81). Since

$$(1 - \tau)\hat{\eta} = \tau\sigma = \hat{\eta}\sigma / (\hat{\eta} + \sigma),$$

we have proved (12.19), but in the interval [-2,2], rather than in $\mathbb{R}$. The boundedness of $P_n(x)W(a_nx)$ in $\mathbb{R}$ ensures that it converges geometrically to 0 in $\mathbb{R}\setminus[-2,2]$.

Finally, if h(x) is positive in [-1,1], we may use the polynomials of Lemma 12.5, much as in the proof of Theorem 12.1, to add a small imaginary part, ensuring that (12.21) is true. □

13. Bernstein's Formula and Bernstein Extremal Polynomials.

In this section, we gather some material on Bernstein's formula, which gives an explicit form for L_p extremal errors and extremal polynomials in the case of certain weights on $[-1,1]$. Combined with the weighted polynomial approximations of section 12, Bernstein's formula will enable us to obtain asymptotics for $E_{np}(W)$. We also recall from [33] some sufficient conditions for asymptotics for $E_{np}(W)$.

Theorem 13.1

Let q *be a positive integer and let* $S(x)$ *be a polynomial of degree* $2q$, *positive in* $(-1,1)$, *possibly with simple zeros at* ± 1, *and let*

(13.1) $$V(x) := \begin{cases} \{(1-x^2)/S(x)\}^{1/2}, & x \in (-1,1), \\ 0, & x \in \mathbb{R}\backslash(-1,1). \end{cases}$$

Further for $0 < p \leq \infty$, *let*

(13.2) $$V_p(x) := \begin{cases} (1-x^2)^{-1/(2p)}\, V(x), & x \in (-1,1), \\ 0, & x \in \mathbb{R}\backslash(-1,1), \end{cases}$$

and let

(13.3) $$\sigma_p := \begin{cases} \{\Gamma(1/2)\,\Gamma((p+1)/2)\,/\,\Gamma(p/2+1)\,\}^{1/p}, & p < \infty, \\ 1, & p = \infty. \end{cases}$$

(a) *Then for* $1 \leq p \leq \infty$ *and* $n \geq q$,

(13.4) $$E_{np}(V_p) = \inf_{P \in \mathcal{P}_{n-1}} \|(x^n - P(x))V_p(x)\|_{L_p[-1,1]}$$

(13.5) $$= \sigma_p\, 2^{-n+1}\, G[V] = \sigma_p\, 2^{-n+1-1/p}\, G[V_p],$$

where the weighted geometric mean $G[.]$ *is defined by* (2.12). *Further for* $0 < p < 1$,

(13.6) $$E_{np}(V_p) \leq \sigma_p\, 2^{-n+1-1/p}\, G[V_p].$$

(b) <u>For</u> $x \in [-1,1]$, <u>write</u>

(13.7) $\qquad x = (z + z^{-1})/2, \; z = e^{i\theta}, \; \theta \in [0,\pi].$

<u>For</u> $n \geq q$, <u>let</u>

(13.8) $\qquad T_n(x) := T_n(x,V),$

$$:= 2^{-n} G[V] \{z^{-n} D^{-2}(V(\cos\phi);z) + z^n D^{-2}(V(\cos\phi);z^{-1})\},$$

<u>where</u> $D(.;z)$ <u>denotes the Szegö function defined by (2.14). Then</u> $T_n(x)$ <u>is a monic polynomial of degree n and for</u> $0 < p < \infty$,

(13.9) $\qquad \int_{-1}^{1} |T_n(x)|^{p-2} T_n(x) x^k V_p^p(x) dx = 0, \; k = 0,1,2,\ldots,n-1,$

<u>while for</u> $0 < p \leq \infty$,

(13.10) $\qquad \|T_n(x)V_p(x)\|_{L_p[-1,1]} = \sigma_p \, 2^{-n+1-1/p} G[V_p].$

<u>Furthermore</u>,

(13.11) $\qquad |T_n(x)V(x)| \leq 2^{-n+1-1/p} G[V_p], \; x \in [-1,1],$

<u>and for</u> $u \in \mathbb{C}\backslash[-1,1]$,

(13.12) $\qquad |T_n(u)/\{ 2^{-n-1/p} G[V_p] \varphi(u)^n D^{-2}(V(\cos\phi); \varphi(u)^{-1}) \} - 1 |$

$$\leq |\varphi(u)|^{2q-2n-2}.$$

(c) <u>For</u> $1 \leq p \leq \infty$, $n \geq q$,

(13.13) $\qquad T_{np}(V_p,x) \equiv T_n(x).$

The statements above appear in Achieser [1,pp. 250-254], at least for $p \geq 1$, but expressed in a slightly different form. It seems likely that equality in (13.6) should hold for $0 < p < 1$.

In proving Theorem 13.1, we shall assume, as does Achieser, that $\delta(0) = 1$. The general case follows by multiplication by a suitable constant.

<u>Background to Theorem 13.1</u>

We first reexpress the statements in Achieser in a suitable form.

Write

$$S(x) = \prod_{k=1}^{2q} (1 - x/a_k),$$

$$a_k =: (c_k + c_k^{-1})/2, \quad |c_k| \le 1, \quad k=1,2,\ldots,2q,$$

$$x = (z + z^{-1})/2, \quad x \in [-1,1], \quad |z| = 1,$$

$$\Omega(z) := \{ \prod_{k=1}^{2q} (z - c_k)\}^{1/2},$$

$$L_{n+1} := 2^{-n}\{ \prod_{k=1}^{2q} (1 + c_k^2)\}^{1/2}.$$

Note that the c_k occur in conjugate pairs, so L_{n+1} is a well defined positive number. Further, the only c_k which can possibly lie on the unit circle are $c_k = \pm 1$. We can then define a branch of $\Omega(z)$ analytic and single-valued in $\{z : |z| \ge 1\}\backslash\{-1,1\}$. Finally, let

$$(13.14) \qquad U_n(x) := L_{n+1}\{z^{2q-n-1} \frac{\Omega(1/z)}{\Omega(z)} - z^{n+1-2q} \frac{\Omega(z)}{\Omega(1/z)}\} \frac{S(x)^{1/2}}{z^{-1} - z}.$$

Achieser [1,p.250] notes that

$$(13.15) \qquad \prod_{k=1}^{2q} (1 + c_k^2) = G[S]^{-1}.$$

This may easily be derived from Jensen's formula. Thus

$$(13.16) \qquad L_{n+1} = 2^{-n} G[S]^{-1/2}.$$

Achieser [1,pp.253-254] shows that for $k = 0,1,\ldots,n-1$ and $p \ge 1$,

$$(13.17) \qquad \int_{-1}^{1} |U_n(x)V(x)|^{p-1} \operatorname{sign}(U_n(x)) \, x^k \, S(x)^{-1/2} \, dx = 0.$$

For $p = 1$, this relation implies that $U_n(x)$ has n simple zeros in $(-1,1)$. Then we see that this last integral converges absolutely for all $p \in \mathbb{C}$ such that $\mathrm{Re}(p) > 0$, and further defines a single-valued analytic function of p for $\mathrm{Re}(p) > 0$. Hence by using analytic continuation, (13.17) is true for all $p > 0$. Using (13.2), we may rewrite this in the form

(13.18) $$\int_{-1}^{1} |U_n(x)|^{p-2} U_n(x) x^k \, V_p^p(x)\, dx = 0,\ k = 0,1,2,\dots,n-1,\ p > 0.$$

Next, Achieser [1,p.254] shows that for $1 \le p < \infty$,

(13.19) $$\int_{-1}^{1} |U_n(x)V(x)|^p \,(1 - x^2)^{-1/2}\, dx = \sigma_p^p \, L_{n+1}^p.$$

By a continuation argument (both sides are analytic in p in a neighbourhood of $(0,\infty)$), (13.19) persists for $0 < p < 1$. Thus for $0 < p < \infty$,

(13.20) $$\|U_n V_p\|_{L_p[-1,1]} = \sigma_p \, L_{n+1}.$$

Finally letting $p \to \infty$, we see that (13.20) persists for $p = \infty$, noting (via standard asymptotics for the gamma function) that $\sigma_p \to 1$, $p \to \infty$. Achieser [1,p.251] also shows that for $1 \le p < \infty$,

(13.21) $$E_{np}(V_p) = \sigma_p \, L_{n+1},$$

and

(13.22) $$T_{np}(V_p,x) \equiv U_n(x).$$

We now show that (13.21) and (13.22) persist for $p = \infty$. In the case when S(x) has simple zeros at ± 1, we remark that this is shown in [1,p.250]. Since $U_n(x)$ is monic [1,p.251], we have from (13.20),

$$E_{n\infty}(V_\infty) \le \|U_n V_\infty\|_{L_\infty[-1,1]} = \sigma_\infty \, L_{n+1}.$$

In the other direction, we note that for each fixed monic polynomial P(x) of degree n,

$$\|PV_\infty\|_{L_\infty[-1,1]} = \lim_{p\to\infty} \|PV_p\|_{L_p[-1,1]}.$$

This follows by a straightforward argument. Hence we deduce that

$$E_{n\infty}(V_\infty) \ge \liminf_{p \to \infty} E_{np}(V_p) = \sigma_\infty \, L_{n+1}.$$

Therefore (13.21) persists and in view of (13.20) for $p = \infty$, so does (13.22).

Proof of (a) of Theorem 13.1

By a standard integral [6,p.243,nos.(864.31) and (864.32)],

(13.23) $G[1 - x^2] = \exp(\ \pi^{-1}\int_{-1}^{1} \log\ (1 - x^2)\ dx\ /\ \sqrt{1 - x^2}) = 1/4.$

Then, using the multiplicative properties of G,

(13.24) $G[V] = (1/2)G[S(x)]^{-1/2},$

and

$$G[V_p] = 2^{-1+1/p}G[S(x)]^{-1/2}.$$

Then from (13.16),

$$L_{n+1} = 2^{-n+1}G[V] = 2^{-n+1-1/p}G[V_p].$$

Then (13.21) yields (13.5). For $0 < p < 1$, we have from (13.20),

$$E_{np}(V_p) \leq \|U_n V_p\|_{L_p[-1,1]} = \sigma_p\ L_{n+1}.$$

Hence (13.6). □

Proof of (b),(c) of Theorem 13.1

If we can show that $U_n(x)$ of (13.14) and $T_n(x)$ of (13.8) are identical; then (13.9) and (13.10) follow from (13.18) and (13.20) respectively. Further (13.22) yields (13.13), while (13.11) follows from (13.10) with $p = \infty$ and (13.5). It thus remains to prove $U_n(x) = T_n(x)$ and then (13.12). Now by definition of $\{c_k\}$ and $\Omega(z)$, and by (13.15) and some elementary manipulations, we see that

(13.25) $\Omega^2(z)\ \Omega^2(1/z) = S(x)\ G[S]^{-1}.$

Letting

$$H(z) := \prod_{k=1}^{2q} (1 - zc_k) = z^{2q}\ \Omega^2(1/z)$$

and bringing the difference in (13.14) to a common denominator, as well as using (13.16), we see that

$$U_n(x) = 2^{-n}\{\ z^{-n-1}H(z) - z^{n+1}H(1/z)\ \}/(z^{-1} - z\}.$$

Further, note that $H(z)$ has no zeros in $|z| < 1$, $H(0) > 0$, and by (13.25),

$$|H(z)|^2 = H(z)H(1/z) = \Omega^2(z)\ \Omega^2(1/z) = G[S]^{-1}\ S(x).$$

Then (10.2.1) in [66,p.275] and (10.2.12) in [66,p.277] show that

$$D(S(\cos\ \phi);z) = H(z)\ G[S]^{1/2}.$$

Further, by (10.2.13) in [66,p.277],

$$D(\sin^2\phi;z) = (1/2)(1 - z^2).$$

Then

$$U_n(x) = 2^{-n-1} G[S]^{-1/2} \{z^{-n} \frac{D(S(\cos\phi);z)}{D(\sin^2\phi;z)} + z^n \frac{D(S(\cos\phi);z^{-1})}{D(\sin^2\phi;z^{-1})}\}$$

$$= 2^{-n-1}G[S]^{-1/2}\{z^{-n}D^{-2}(V(\cos\phi);z) + z^n D^{-2}(V(\cos\phi);z^{-1})\},$$

by multiplicative properties of D. Finally, (13.24) and (13.8) yield $U_n(x) = T_n(x)$.

To prove (13.12), we note that $z^{-2q}H(z)/H(1/z)$ is a rational function analytic in $|z| \geq 1$, with unit modulus on $|z| = 1$. Thus for $u \in \mathbb{C}\backslash[-1,1]$,

$$| \varphi(u)^{-2q} H(\varphi(u))/H(1/\varphi(u)) | \leq 1,$$

and easy manipulations yield (13.12). □

Next, we recall some sufficient conditions for asymptotic upper and lower bounds for $E_{np}(W)$, derived in [33], using Bernstein's formula, Hölder's inequality, and some elementary manipulations.

Proposition 13.2

Let $1 \leq p < \infty$, and let $W(x)$ be a non-negative function such that for every $q \in [p,\infty)$,

$$x^n W(x) \in L_q(\mathbb{R}),\ n=0,1,2,\ldots .$$

Assume further that there exist respectively increasing and decreasing sequences $\{c_n\}_1^\infty$ and $\{\delta_n\}_1^\infty$ of positive numbers, such that

(13.26) $$\lim_{n\to\infty} \delta_n = 0,$$

and for $n = 1,2,3,\ldots$, and each $P \in \mathcal{P}_n$,

(13.27) $$\|PW\|_{L_p(\mathbb{R})} \leq (1 + \delta_n)\|PW\|_{L_p(-c_n,c_n)}.$$

Finally, assume that for every $q \in [p,\infty)$ and each $g(x)$ positive and continuous in [-1,1], there exists $P_{2n-2} \in \mathcal{P}_{2n-2}$, positive in [-1,1], n=1,2,3,..., such that

$$(13.28) \qquad \liminf_{n \to \infty} \int_{-1}^{1} \log \{ \sqrt{P_{2n-2}(x)}\, W(c_n x)\, g(x) \}\, dx / \sqrt{1 - x^2} \geq 0,$$

and

$$(13.29) \qquad \limsup_{n \to \infty} \| \sqrt{P_{2n-2}(x)}\, W(c_n x)\, g(x) \|_{L_q[-1,1]} \leq 2^{1/q}.$$

Then, if σ_p is given by (13.3),

$$(13.30) \qquad \limsup_{n \to \infty} E_{np}(W) / \{(c_n/2)^{n+1/p}\, G[W(c_n x)]\} \leq 2\sigma_p,$$

where $G[\cdot]$ is defined as in (2.13).

The above is Proposition 2.1 in [33], except that we have removed a redundant condition (namely (2.1) in [33] -see the comment there after the proposition). In the next section we shall establish a slightly different condition. Following is Proposition 2.2 in [33]: Again, we have dropped (2.1) in [33], noting that it is implied by (13.31) below.

Proposition 13.3

Let $1 < p < \infty$, and let $W(x)$ be a non-negative function such that

$$x^n W(x) \in L_p(\mathbb{R}),\ n = 0,1,2,\dots,$$

and such that for $q \in (p,\infty)$ and all $a > 0$,

$$(13.31) \qquad W(x)^{-1} \in L_q[-a,a].$$

Assume further that $\{d_n\}_1^\infty$ is an increasing sequence of positive numbers with the following property: For every $q \in (p,\infty)$ and each $g(x)$ even, positive and continuous in [-1,1], there exists $P_{2n} \in \mathcal{P}_{2n}$, positive in [-1,1], n=1,2,3,..., such that

$$(13.32) \qquad \limsup_{n \to \infty} \int_{-1}^{1} \log \{ \sqrt{P_{2n}(x)}\, W(d_n x)\, g(x) \}\, dx / \sqrt{1 - x^2} \leq 0,$$

and

(13.33) $\limsup_{n\to\infty} \| \{\sqrt{P_{2n}(x)}\, W(d_n x)\, g(x)\}^{-1} \|_{L_q[-1,1]} \le 2^{1/q}$.

Then

(13.34) $\liminf_{n\to\infty} E_{np}(W) / \{(d_n/2)^{n+1/p}\, G[W(d_n x)]\} \ge 2\sigma_p$.

We remark that the integrability conditions on W in Proposition 13.2 (resp. Proposition 13.3) allow zeros (poles) of algebraic strength, but only weak, for example logarithmic, poles(zeros). Even for weights on [-1,1], the authors [34] have been unable to circumvent this restriction. Of course for p = 2, there are the special methods for orthogonal polynomials. We shall use these in Section 16.

In applications, the sequences $\{c_n\}_1^\infty$ and $\{d_n\}_1^\infty$ are different, but sufficiently close to deduce asymptotics for $E_{np}(W)$, with the aid of the following lemma, which is Lemma 2.3 in [33].

Lemma 13.4

Let $W(x) := e^{-Q(x)}$, where $Q(x)$ is even, continuous in $\mathbb{R}$ and $Q''(x)$ exists for $x > 0$, while $xQ'(x)$ is positive and increasing in $(0,\infty)$ with

(13.35) $\lim_{|x|\to\infty} xQ'(x) = \infty$.

Assume further that there exist $C_3, C_4 > 0$ such that

(13.36) $x|Q''(x)|/Q'(x) \le C_3, \quad x \in (0,\infty)$

and

(13.37) $Q'(2x)/Q'(x) \le C_4, \quad x \in (0,\infty)$.

Let $a_n = a_n(W)$, n large enough, and let $\{e_n\}_1^\infty$ be a sequence of real numbers satisfying

(13.38) $\lim_{n\to\infty} n^{1/2}(e_n/a_n - 1) = 0$.

Then for $0 < p \le \infty$,

$$\lim_{n\to\infty} e_n^{n+1/p}\, G[W(e_n x)] \,/\{\, a_n^{n+1/p}\, G[W(a_n x)] \,\} = 1. \tag{13.39}$$

We note that the proof of Lemma 13.4 given in [33] involves a simple Taylor expansion of $Q(e_n t) = \log W(e_n t)$ in terms of $Q(a_n t)$.

14. Proof of The Asymptotics for $E_{np}(W)$.

In this section, we prove the asymptotics for $E_{np}(W)$ stated in Theorems 3.1 and 3.2. First, we prove the following generalization of (3.5) in Theorem 3.1.

Theorem 14.1

Let $W(x) := e^{-Q(x)}$, *where* $Q(x)$ *is even and continuous in* $\mathbb{R}$, $Q''(x)$ *exists and is continuous in* $(0,\infty)$, *and* $Q'(x)$ *is positive in* $(0,\infty)$, *while for some* $C_1, C_2 > 0$,

$$(14.1) \qquad C_1 \le (xQ'(x))'/Q'(x) \le C_2, \quad x \in (0,\infty).$$

Suppose, further, that $Q'''(x)$ *exists and is continuous for* x *large enough, with*

$$(14.2) \qquad x^2|Q'''(x)|/Q'(x) \le C_3, \qquad x \text{ large enough}.$$

Let $a_n = a_n(W)$ *for* n *large enough. Let* $0 < p < \infty$, *and let* $\hat{W}(x)$ *be a non-negative function such that*

$$(14.3) \qquad \hat{W}(x) \in L_p(\mathbb{R}),$$

and

$$(14.4) \qquad \limsup_{|x|\to\infty} \hat{W}(x)/W(x) \le 1.$$

Then if σ_p *and* $G[\cdot]$ *are as in* (2.17) *and* (2.13) *respectively,*

$$(14.5) \qquad \limsup_{n\to\infty} E_{np}(\hat{W})/\{ (a_n/2)^{n+1/p} G[W(a_n x)] \} \le 2\sigma_p.$$

The essential feature of the above result is that the behaviour of $\hat{W}$ in a fixed finite interval is irrelevant for the asymptotic upper bound (14.5).

Proof of Theorem 14.1

We first estimate the L_p norm of weighted polynomials for $\hat{W}$ in terms of those for W. Let $\epsilon > 0$. We can choose $C_4 > 0$, such that

$$\hat{W}(x) \le (1+\epsilon)W(x), \quad |x| \ge C_4.$$

Let $K > 0$ be a sufficiently large positive number and

$$(14.6) \qquad e_n := a_n(1 + K((\log n)/n)^{2/3}), \text{ n large enough,}$$

Then, by Theorem 7.2, we have for $P \in \mathcal{P}_n$,

$$\|P\hat{W}\|_{L_p(|x|\ge C_4)} \le (1+\epsilon)(1+n^{-C_5})\|PW\|_{L_p(|x|\le e_n)}.$$

Making the substitution $x = e_n u$ yields

$$\|P\hat{W}\|_{L_p(|x|\ge C_4)} \le e_n^{1/p}(1+\epsilon)(1+n^{-C_5})\|(PW)(e_n u)\|_{L_p[-1,1]}.$$

Then

$$(14.7) \qquad E_{np}(\hat{W})^p = \inf_{P \in \mathcal{P}_{n-1}} \{\|(x^n - P(x))\hat{W}(x)\|^p_{L_p(\mathbb{R})}\}$$

$$\le e_n^{np+1} \inf_{P \in \mathcal{P}_{n-1}} \{ \|(u^n - P(u))\hat{W}(e_n u)\|^p_{L_p(|u|\le C_4/e_n)}$$

$$+ (1+\epsilon)^p(1+n^{-C_5})^p\|(u^n - P(u))W(e_n u)\|^p_{L_p[-1,1]}\}.$$

Now let $S(u) := (1-u^2)\hat{S}(u)$, where $\hat{S}(u)$ is a polynomial of degree at most 2n-2, positive in [-1,1], and let V and V_p be given by (13.1) and (13.2) respectively. For notational simplicity, we shall suppress the dependence of S,V and V_p on n. Inserting $T_n(x)$, defined by (13.8), into (14.7), and using (13.10) and (13.11), we obtain for n large enough,

$$(14.8) \qquad E_{np}(\hat{W})^p \le e_n^{np+1} \{ \|T_n(u)\hat{W}(e_n u)\|^p_{L_p(|u|\le C_4/e_n)}$$

$$+ (1+2\epsilon)^p \|T_n(u)V_p(u)\|^p_{L_p[-1,1]} \|V_p^{-1}(u)W(e_n u)\|^p_{L_\infty(\epsilon\le|u|\le1)}$$

$$+ (1+2\epsilon)^p \|T_n(u)W(e_n u)\|^p_{L_p(|u|\le\epsilon)} \}$$

$$\le (e_n/2)^{np+1}(2\sigma_p)^p\, G[V_p]^p \{\sigma_p^{-p} \|V^{-1}(u)\hat{W}(e_n u)\|^p_{L_p(|u|\le C_4/e_n)}$$

$$+ (1+2\epsilon)^p \|V_p^{-1}(u)W(e_n u)\|^p_{L_\infty(\epsilon\le|u|\le1)}$$

$$+ (1+2\epsilon)^p \sigma_p^{-p} \|V^{-1}(u)W(e_n u)\|^p_{L_p(|u|\le\epsilon)} \}.$$

Here, by choice of S,

$$V^{-1}(u)W(e_n u) = \hat{S}^{1/2}(u)W(e_n u),$$

and

$$V_p^{-1}(u)W(e_n u) = (1 - u^2)^{1/(2p)}\,\hat{S}^{1/2}(u)W(e_n u).$$

We now choose the polynomial $\hat{S}$ using Theorem 12.1. Let g(x) be positive and continuous in [-1,1].

Then, since (14.6) holds, Theorem 12.1 shows that we can choose polynomials $P_{n-1}(u)$ of degree at most n-1 such that

$$(14.9)\qquad \|P_{n-1}(u)W(e_n u)\|_{L_\infty(\mathbb{R})} \le C_6,\ n = 1,2,3,\ldots\ ,$$

$$(14.10)\qquad \lim_{n\to\infty} P_{n-1}(u)W(e_n u) = g(u),$$

uniformly in compact subsets of {x: 0 < |x| <1}, and

$$(14.11)\qquad \lim_{n\to\infty} \int_{-1}^{1} \log |P_{n-1}(u)W(e_n u)|\,(1 - u^2)^{-1/2}\,du$$

$$= \int_{-1}^{1} \log g(u)\,(1 - u^2)^{-1/2}\,du.$$

Let $\hat{S} := \hat{S}_n := |P_{n-1}|^2 = \mathrm{Re}(P_{n-1})^2 + \mathrm{Im}(P_{n-1})^2$, a polynomial of degree at most 2n-2, non-negative in [-1,1]. We may assume $\hat{S}$ is positive in [-1,1], by adding, if necessary, a small positive number. Then, by (14.11), and (13.2),

$$(14.12)\qquad G[V_p]\,G[W(e_n u)]^{-1}\,G[(1 - x^2)^{1/(2p)}g(x)] \to 1, \qquad n \to \infty.$$

Note too that since (13.38) is satisfied, and since (14.1) implies (13.35) and (13.36), while (6.19) implies (13.37), Lemma 13.4 shows that

$$(14.13)\qquad \lim_{n\to\infty} e_n^{n+1/p}\,G[W(e_n u)]\,/\{a_n^{n+1/p}\,G[W(a_n u)]\} = 1.$$

Next, if $0 < \delta < 1$, we have from (14.10),

$$\|V_p^{-1}(u)W(e_n u)\|_{L_\infty(\epsilon\le|u|\le 1-\delta)} \to \|(1 - u^2)^{1/(2p)}g(u)\|_{L_\infty(\epsilon\le|u|\le 1-\delta)},$$

$n \to \infty$, while from (14.9),

$$\|V_p^{-1}(u)W(e_n u)\|_{L_\infty(1-\delta\le|u|\le 1)} \le$$

$$\leq (1 - (1 - \delta)^2)^{1/(2p)} \ \|P_{n-1}(u)W(e_n u)\|_{L_\infty(1-\delta\leq|u|\leq 1)}$$

$$\leq C_6 \ (3\delta)^{1/(2p)} < \epsilon,$$

if δ is small enough. Hence for n large enough,

$$\|V_p^{-1}(u)W(e_n u)\|_{L_\infty(\epsilon\leq|u|\leq 1)} \leq \|(1 - u^2)^{1/(2p)} g(u)\|_{L_\infty[-1,1]} + \epsilon.$$

Substituting this last inequality and (14.9),(14.12) and (14.13) into the extreme right-hand side of (14.8), we obtain

$$E_{np}(\hat{W})/\{ (a_n/2)^{n+1/p} \ G[W(a_n x)] \ 2\sigma_p \}$$

$$\leq (1 + o(1)) \ G[(1 - x^2)^{1/(2p)} g(x)]^{-1}$$

$$\times \{ \sigma_p^{-p} \ C_6^p \ \|W^{-1}(e_n u)\hat{W}(e_n u)\|^p_{L_p(|u|\leq C_4/e_n)}$$

$$+ (1 + 2\epsilon)^p (\ \|(1 - u^2)^{1/(2p)} g(u)\|_{L_\infty[-1,1]} + \epsilon \ )^p$$

$$+ (1 + 2\epsilon)^p \ \sigma_p^{-p} \ C_6^p \ 2\epsilon \ \}^{1/p}.$$

Here by a substitution, we see that

$$\|W^{-1}(e_n u)\hat{W}(e_n u)\|_{L_p(|u|\leq C_4/e_n)} = O(e_n^{-1/p}) \to 0, \ n \to \infty.$$

Choosing g(u) to approximate $(1 - u^2)^{-1/(2p)}$ in the sense that

$$G[(1 - u^2)^{1/(2p)} g(u)]^{-1} \leq 1 + \epsilon,$$

and

$$\|(1 - u^2)^{1/(2p)} g(u)\|_{L_\infty[-1,1]} \leq 1 + \epsilon,$$

and then letting $\epsilon \to 0$, yields (14.5). □

<u>Proof of (3.4) and (3.5) in Theorem 3.1</u>

The asymptotic upper bound (3.5) follows for all $0 < p < \infty$ from Theorem 14.1. We proceed to prove a corresponding asymptotic lower bound for $E_{np}(W)$, $1 < p < \infty$. Let

$$\epsilon_n := -(\log \log n) \ ((\log n)/n)^{2/3}, \ n = 2,3,4,\ldots ,$$

and

$$e_n := a_n(1 + \epsilon_n), \ n = 2,3,4,\ldots .$$

Then we have

$$\lim_{n\to\infty} \epsilon_n \{n/\log n\}^{2/3} = -\infty.$$

Then (12.4),(12.5) and (12.11) are satisfied. By Theorem 12.1, more precisely by (12.7),(12.8) and (12.13), we may find polynomials $P_n(x)$ of degree at most n, such that

$$C_5 \geq |P_n(x)W(e_n x)| \geq C_6, \; x \in [-1,1], \; n = 1,2,3,\ldots ,$$

and

$$\lim_{n\to\infty} P_n(x)W(e_n x) = g^{-1}(x),$$

uniformly in compact subsets of $\{x: 0 < |x| < 1\}$, where $g(x)$ is a given function continuous and positive in $[-1,1]$. Then setting

$$S_{2n}(x) := |P_n(x)|^2 = \mathrm{Re}(P_n)^2 + \mathrm{Im}(P_n)^2,$$

we see that S_{2n} is a polynomial positive in $[-1,1]$ and for $q \in [p,\infty)$, $1 < p < \infty$,

$$\lim_{n\to\infty} \|\{S_{2n}^{1/2}(x)W(e_n x)g(x)\}^{-1}\|_{L_q[-1,1]} = \|1\|_{L_q[-1,1]} = 2^{1/q},$$

and

$$\lim_{n\to\infty} \int_{-1}^{1} \log\, \{S_{2n}^{1/2}(x)W(e_n x)g(x)\}\, (1 - x^2)^{-1/2}\, dx = 0.$$

By Proposition 13.3, for $1 < p < \infty$,

$$\liminf_{n\to\infty} E_{np}(W)/\{\, (e_n/2)^{n+1/p}\, G[W(e_n x)]\, \} \geq 2\sigma_p.$$

Since

$$e_n/a_n = 1 + o(n^{-1/2}), \; n \to \infty,$$

Lemma 13.4 yields the required lower bound

$$\liminf_{n\to\infty} E_{np}(W)/\{\, (a_n/2)^{n+1/p}\, G[W(a_n x)]\, \} \geq 2\sigma_p. \;\square$$

Next, we deal with the cases $p = 1,\infty$:

<u>Proof of the Asymptotics for</u> $E_{np}(W)$, $p = 1,\infty$, <u>in Theorem 3.2.</u>
We first establish asymptotic lower bounds for $E_{np}(W)$, $p = 1,\infty$. Under our assumption (3.12),

$$\theta_n := a_n/n = o(n^{-1/2}), \; n \to \infty.$$

Then we can find a sequence $\{r_n\}_1^\infty$ satisfying

$$\lim_{n\to\infty} r_n = \infty,$$

and

$$\theta_n r_n = o(n^{-1/2}), \ n \to \infty.$$

By Theorem 12.2, there exist polynomials $P_n(x)$ of degree at most n-1 such that

(14.14) $$\lim_{n\to\infty} \|1 - P_n(x)W(a_n x)\|_{L_\infty(|x|\le 1-r_n(\theta_n+\rho_n))} = 0,$$

where

$$\rho_n := ((\log n)/n)^{2/3}, \ n = 2,3,4,\dots .$$

We may also assume that $\{r_n\}_1^\infty$ grows so slowly that

$$\eta_n := -r_n(\theta_n + \rho_n), \ n = 2,3,4,\dots,$$

satisfies

(14.15) $$\eta_n = o(n^{-1/2}), \ n \to \infty.$$

Now set

(14.16) $$e_n := a_n(1 + \eta_n), \ n = 2,3,4,\dots ,$$

and

$$\hat{P}_n(u) := P_n(ue_n/a_n), \ n = 2,3,4,\dots .$$

We have from (14.14),

(14.17) $$\lim_{n\to\infty} \|1 - \hat{P}_n(u)W(e_n u)\|_{L_\infty[-1,1]} = 0.$$

Next, if $p = \infty$, let us set $\hat{S}_{2n-2}(u) := |\hat{P}_n(u)|^2$, a positive polynomial of degree at most 2n-2, and let $S(u) := (1 - u^2)\hat{S}_{2n-2}(u)$, so that if $V(u)$ and $V_p(u)$ are given by (13.1) and (13.2) respectively, $V_\infty(u) = V(u) = |\hat{P}_n(u)|^{-1}$.

If $p = 1$, we set $S(u) := \hat{S}_{2n-2}(u)$, so that $V_1(u) = |\hat{P}_n(u)|^{-1}$. Thus for either $p = 1$ or $p = \infty$, we can choose $V_p = |\hat{P}_n|^{-1}$. Then, for $p = 1,\infty$,

$$E_{np}(W) \ge e_n^{n+1/p} \min_{P \in \mathcal{P}_{n-1}} \|(u^n - P(u))W(e_n u)\|_{L_p[-1,1]}$$

$$\ge e_n^{n+1/p} \min_{P\in\mathcal{P}_{n-1}} \|(u^n-P(u))V_p(u)\|_{L_p[-1,1]} \min_{[-1,1]} |W(e_n u)V_p^{-1}(u)|$$

$$\geq (e_n/2)^{n+1/p}\ 2\sigma_p\ G[V_p]\ (1 + o(1))$$

$$= (e_n/2)^{n+1/p}\ 2\sigma_p\ G[W(e_n u)]\ (1 + o(1)),$$

by (14.17) and the fact that $G[1] = 1$. Finally, (14.15),(14.16) ensure that

$$e_n/a_n = 1 + o(n^{-1/2}),\ n \to \infty.$$

Then Lemma 13.4 yields for $p = 1,\infty$

$$\text{(14.18)} \qquad \liminf_{n \to \infty} E_{np}(W)/\{\ (a_n/2)^{n+1/p}\ G[W(a_n u)]\ \} \geq 2\sigma_p.$$

In the case when $H := 1/W$ is an even entire function with non-negative Maclaurin series coefficients, we can use (11.10) in Theorem 11.2, with $\eta = 0$, $g = 1$, to obtain polynomials $\{P_n\}_1^\infty$ satisfying (14.14).

It remains to prove the asymptotic upper bound when $p = \infty$: For $p = 1$, Theorem 14.1 yields the asymptotic upper bound. Let $h(t)$ be a polynomial positive in $[-1,1]$, and

$$g(t) := \begin{cases} (1 - t^2)\ h(t), & t \in [-1,1], \\ 0 & , t \in \mathbb{R}\backslash[-1,1]. \end{cases}$$

By Theorem 12.3, we can choose polynomials $P_n(x)$ of degree at most $n - 1$ such that

$$\text{(14.19)} \qquad \lim_{n\to\infty} \|g(x) - P_n(x)W(a_n x)\|_{L_\infty(\mathbb{R})} = 0,$$

and

$$\text{(14.20)} \qquad \lim_{n\to\infty} \int_{-1}^{1} \frac{\log |P_n(x)W(a_n x)|}{(1 - x^2)^{1/2}}\, dx = \int_{-1}^{1} \frac{\log g(x)}{(1 - x^2)^{1/2}}\, dx.$$

Let us set $\hat{S}_{2n-2}(x) := |P_n(x)|^2$, a polynomial of degree at most $2n - 2$, non-negative in $[-1,1]$. We can ensure that $\hat{S}_{2n-2}(x)$ is positive in $[-1,1]$, by adding if necessary, a small positive number. Let $S_{2n}(x) := (1 - x^2)\hat{S}_{2n-2}(x)$, so that V_∞ of (13.2) satisfies $V_\infty(x) = |P_n(x)|^{-1}$. Then using (7.3) in Theorem 7.1, we have

$$E_{n\infty}(W) = a_n^n \min_{P \in \mathcal{P}_{n-1}} \|(u^n - P(u))W(a_n u)\|_{L_\infty[-1,1]}$$

$$\leq a_n^n \min_{P \in \mathcal{P}_{n-1}} \|(u^n - P(u))V_\infty(u)\|_{L_\infty[-1,1]} \|V_\infty^{-1}(u)W(a_n u)\|_{L_\infty[-1,1]}$$

$$\leq (a_n/2)^n \, 2\sigma_\infty \, G[V_\infty] \, (\|g\|_{L_\infty(\mathbb{R})} + o(1))$$

$$\leq (a_n/2)^n \, 2\sigma_\infty \, G[W(a_n u)] \, G[g]^{-1} \, (\|g\|_{L_\infty(\mathbb{R})} + o(1)),$$

by (13.5),(14.19), and (14.20). Choosing g close to 1 in a suitable sense then yields the asymptotic upper bound corresponding to (14.18). Finally, if $H := 1/W$ is an even entire function with non-negative Maclaurin series coefficients, we can similarly use Theorem 11.2 and Lemma 12.5 to produce polynomials satisfying (14.19) and (14.20). □

15. Proof of the Asymptotics for the L_p Extremal Polynomials.

Our main tool in establishing asymptotics in the plane of L_p extremal polynomials, $2 \le p < \infty$, is the following comparison theorem:

Theorem 15.1

Let W *be a non-negative function such that for some given*

(15.1) $\quad 2 \le p < \infty,$

and for $n = 0,1,2,\ldots$,

(15.2) $\quad x^n W(x) \in L_p(\mathbb{R}).$

Assume further that there exists an increasing sequence $\{c_n\}_1^\infty$ *of positive numbers such that for* $n = 1,2,3,\ldots$ *and* $P \in \mathcal{P}_{n+1}$,

(15.3) $\quad \|P_n W\|_{L_p(\mathbb{R})} \le 2\, \|P_n W\|_{L_p(-c_n,c_n)}.$

Let

$$P(x) := Ax^n + \ldots \in \mathcal{P}_n,\ A \ne 0,$$

and let

(15.4) $\quad r_n := A\, E_{np}(W).$

Let $T_{np}(W,x)$ *and* $p_{np}(W,x)$ *be as in* (2.3) *and* (2.4), *and denote their zeros by* $-\infty < x_{nn} < x_{n-1,n} < \ldots < x_{1n} < \infty$. *Let*

(15.5) $\quad d_n(z) := \min\{|z/c_n - x_{jn}/c_n| : j=1,2,\ldots,n\},\ n = 1,2,\ldots\ .$

Then for $z \in \mathbb{C}\backslash\{x_{jn}\}_1^n$,

(15.6) $\quad |P(z)/p_{np}(W,z) - r_n| \le 2\,\{\ \|PW\|^2_{L_p(\mathbb{R})} - r_n^2\ \}^{1/2}/d_n(z),$

and

(15.7) $\quad |P(z)/(AT_{np}(W,z)) - 1| \le 2\,\{\ \|PW\|^2_{L_p(\mathbb{R})} r_n^{-2} - 1\ \}^{1/2}/d_n(z).$

Proof

We modify a standard L_2 technique, as in [34] and as in the proof of

Lemma 8.6. Recall from the proof of Lemma 8.6 that $p_{np}(x) := p_{np}(W,x)$ is the orthonormal polynomial of degree n for $d\beta(x) := |p_{np}(W,x)|^{p-2} W^p(x)\,dx$. Let $l_{1n}(x), l_{2n}(x),\ldots,l_{nn}(x)$ denote the fundamental polynomials of Lagrange interpolation at the zeros of $p_{np}(x)$. Recall also from Lemma 8.6 that

$$l_{jn}(x) = \lambda_{jn} p_{n-1}(x_{jn})(\gamma_{n-1}/\gamma_n) p_{np}(x)/(x-x_{jn}),\ j = 1,2,\ldots,n. \tag{15.8}$$

Here the λ_{jn} are the Christoffel numbers of order n for $d\beta(x)$, $p_{n-1}(x)$ is the orthonormal polynomial of degree n-1 for $d\beta(x)$, and γ_{n-1} and γ_n are the leading coefficients of $p_{n-1}(x)$ and $p_{np}(x)$ respectively. Proceeding exactly as in the proof of Lemma 8.6, we obtain as in (8.16),

$$\gamma_{n-1}/\gamma_n \le 2c_n, \tag{15.9}$$

for $n \ge 1$. Now let

$$q(x) := P(x) - r_n p_{np}(W,x),$$

a polynomial of degree at most n-1. We have

$$\begin{aligned} |P(z)/p_{np}(W,z) - r_n| &= |q(z)/p_{np}(W,z)| \\ &= \Big|\sum_{j=1}^{n} q(x_{jn})\, l_{jn}(z)\Big| / |p_{np}(W,z)| \\ &\le 2\, d_n(z)^{-1} \sum_{j=1}^{n} \lambda_{jn}\, |p_{n-1}(x_{jn})|\, |q(x_{jn})|, \end{aligned}$$

by (15.8) and (15.9). Using the Cauchy-Schwarz inequality, we see that the sum in this last right-hand side is bounded above by

$$\Big\{\sum_{j=1}^{n} \lambda_{jn}\, p^2_{n-1}(x_{jn})\Big\}^{1/2} \Big\{\sum_{j=1}^{n} \lambda_{jn}\, q^2(x_{jn})\Big\}^{1/2}$$

$$= \Big\{ \int_{-\infty}^{\infty} q^2(x)\, d\beta(x) \Big\}^{1/2}$$

$$= \Big\{ \int_{-\infty}^{\infty} P^2(x)\, d\beta(x) - 2r_n \int_{-\infty}^{\infty} P(x)\, p_{np}(W,x)\, d\beta(x) + r_n^2 \Big\}^{1/2}$$

$$= \Big\{ \int_{-\infty}^{\infty} P^2(x)\, d\beta(x) - r_n^2 \Big\}^{1/2}$$

$$\leq \{ \|PW\|^2_{L_p(\mathbb{R})} \|p_{np}(W,x)W(x)\|^{p-2}_{L_p(\mathbb{R})} - r_n^2 \}^{1/2},$$

where we have used the definition of $d\beta(x)$ and Holder's inequality with indices $2/p$ and $(p - 2)/p$. Then (15.6), and hence also (15.7), follow. □

Proof of (3.7) and (3.8) of Theorem 3.1

First note that by Theorem 7.2, if K is some fixed large positive number, and

$$c_n := a_n(1 + K((\log n)/n)^{2/3}), \; n = 1,2,3,\dots ,$$

we have for n large enough and $P \in \mathcal{P}_n$,

(15.10) $$\|PW\|_{L_p(\mathbb{R})} \leq (1 + n^{-C_3}) \|PW\|_{L_p(-c_n,c_n)}.$$

Now let $\hat{S}(x)$ be a polynomial of degree at most $2n-2$, positive in $[-1,1]$, and let $S(x) := (1 - x^2)\hat{S}(x)$, so that $V(x)$ of (13.1) satisfies $V(x) = [\hat{S}(x)]^{-1/2}$. Further, let $T_n(x)$ be given by (13.8), and

(15.11) $$P(x) := a_n^{-1/p} E_{np}(V_p)^{-1} T_n(x/a_n),$$

a polynomial of degree n, with leading coefficient

$$A := a_n^{-n-1/p} E_{np}(V_p)^{-1}.$$

If r_n denotes the ratio of leading coefficients of $P(x)$ and $p_{np}(W,x)$, then by (3.4) and (13.5),

(15.12) $$\begin{aligned} r_n &= AE_{np}(W) \\ &= E_{np}(V_p)^{-1} 2^{-n-1/p} G[W(a_nx)] (2\sigma_p) (1 + o(1)) \\ &= G[V_p]^{-1} G[W(a_nx)] (1 + o(1)). \end{aligned}$$

Substituting P into (15.7) of Theorem 15.1 yields

(15.13) $$|a_n^n T_n(z/a_n)/T_{np}(W,z) - 1| \leq 2 \{ \|PW\|^2_{L_p(\mathbb{R})} r_n^{-2} - 1\}^{1/2}/d_n(z).$$

Here by (15.10) and a substitution, and if $0 < \epsilon < 1/2$,

(15.14) $$\begin{aligned} \|PW\|_{L_p(\mathbb{R})} &\leq (1 + o(1)) \|PW\|_{L_p(-c_n,c_n)} \\ &\leq (1 + o(1))E_{np}(V_p)^{-1}\|T_n(x)W(a_nx)\|_{L_p(-c_n/a_n,c_n/a_n)} \end{aligned}$$

$$\leq (1+o(1))E_{np}(V_p)^{-1}\{\|V(x)^{-1}W(a_nx)\|_{L_p(-\epsilon,\epsilon)}\sigma_p^{-1}E_{np}(V_p)$$

$$+ \|T_n(x)V_p(x)\|_{L_p(\epsilon\leq|x|\leq 1-\epsilon)}\|V_p(x)^{-1}W(a_nx)\|_{L_\infty(\epsilon\leq|x|\leq 1-\epsilon)}$$

$$+ \|V(x)^{-1}W(a_nx)\|_{L_\infty[-1,1]}(2(c_n/a_n - (1-\epsilon))^{1/p})\sigma_p^{-1}E_{np}(V_p)\}$$

$$\leq (1+o(1))\{\sigma_p^{-1}\|\hat{S}(x)^{1/2}W(a_nx)\|_{L_\infty[-1,1]}2(\epsilon^{1/p}+(c_n/a_n+\epsilon-1)^{1/p})$$

$$+ \|(1 - x^2)^{1/(2p)}\ \hat{S}(x)^{1/2}\ W(a_nx)\|_{L_\infty(\epsilon\leq|x|\leq 1-\epsilon)}\},$$

by (13.5) and (13.11). We have also used the following consequence of (7.3) applied to W^2, rather than W:

$$\|V(x)^{-1}W(a_nx)\|_{L_\infty(\mathbb{R})} = \|V(x)^{-1}W(a_nx)\|_{L_\infty[-1,1]}.$$

Now we choose $\hat{S}(x)$ using Theorem 12.1. Let $g(x)$ be positive and continuous in $[-1,1]$ and let $P_n(x)$ denote a sequence of polynomials of degree at most $n - \log n$, $n = 1,2,3,...,$ satisfying (12.7) to (12.10) with $c_n = a_n$. Let $\hat{S}(x) := \hat{S}_n(x) := |P_n(x)|^2 = \mathrm{Re}(P_n)^2 + \mathrm{Im}(P_n)^2$, a polynomial of degree at most $2n - 2\log n$. Adding, if necessary, a small positive constant, we may ensure that $\hat{S}_n(x)$ is positive in $[-1,1]$. In terms of $\hat{S}_n(x)$, (12.7) to (12.10) become:

$$\|\hat{S}_n(x)^{1/2}W(a_nx)\|_{L_\infty[-1,1]} \leq C_3,$$

(15.15) $$\lim_{n\to\infty} \hat{S}_n(x)^{1/2}W(a_nx) = g(x),$$

uniformly in compact subsets of $\{x: 0 < |x| < 1\}$, and

(15.16) $$\lim_{n\to\infty} \int_{-1}^{1} \frac{|\log \{\hat{S}_n(x)^{1/2}W(a_nx)/g(x)\}|}{(1 - x^2)^{1/2}}\, dx = 0.$$

Then from (15.12) and (15.16),

(15.17) $$r_n = G[(1 - x^2)^{1/(2p)}g(x)]\ (1 + o(1)).$$

Then also from the last inequality in (15.14),

$$\|PW\|_{L_p(\mathbb{R})} \leq (1+o(1))\{6C_3\sigma_p^{-1}\epsilon^{1/p}+\|(1-x^2)^{1/(2p)}g(x)\|_{L_\infty[-1,1]}\}.$$

In view of (8.2) and(15.13), we have for n large enough, with $z = a_nu$, $u \in \mathbb{C}\backslash[-1,1]$,

$$(15.18)\quad |a_n^n\, T_n(u)/T_{np}(W,a_n u) - 1|$$
$$\le 2\{ (C_4\epsilon^{1/p} + \|(1 - x^2)^{1/(2p)} g(x)\|_{L_\infty[-1,1]})^2$$
$$\times\, G[(1 - x^2)^{1/(2p)} g(x)]^{-2}\, (1 + o(1)) - 1\}^{1/2}/d(u,[-1,1]).$$

The inequality clearly also holds uniformly for u in a closed subset of $\mathbb{C}\backslash[-1,1]$. Next, from (13.12) in Theorem 13.1,

$$(15.19)\quad |T_n(u)/\{ 2^{-n-1/p} G[V_p]\, \varphi(u)^n\, D^{-2}(V(\cos\phi);\varphi(u)^{-1}) \} - 1|$$
$$\le |\varphi(u)|^{-2\log n} \to 0,\ n \to \infty,$$

uniformly in closed subsets of $\mathbb{C}\backslash[-1,1]$. Here, as at (15.17),

$$G[V_p] = G[W(a_n x)]\, G[(1 - x^2)^{1/(2p)}\, g(x)]^{-1}\, (1 + o(1)),$$

and from the definition of the Szegö function, and (15.15) and (15.16), we see that uniformly in compact subsets of $|z| < 1$,

$$D^{-2}(V(\cos\phi);z) = D^{-2}(W(a_n \cos\phi) g^{-1}(\cos\phi);z)\, (1 + o(1)).$$

Choosing g to approximate $(1 - x^2)^{-1/(2p)}$ in the sense that

$$\|(1 - x^2)^{1/(2p)} g(x)\|_{L_\infty[-1,1]} \le 1,$$

and for some small $\eta > 0$,

$$\int_{-1}^{1} \frac{|\log \{g(x)(1 - x^2)^{1/(2p)}\}|}{(1 - x^2)^{1/2}}\, dx < \eta,$$

we see that the right-hand side of (15.18) may be made arbitrarily small for u in a closed subset of $\mathbb{C}\backslash[-1,1]$. Together with (15.19), the above inequalities then yield (3.7) and hence also (3.8). □

To handle the case $p = \infty$, we require a uniform in p form of Theorem 15.1:

<u>Lemma 15.2</u>

<u>Let</u> $W(x)$ <u>be a continuous non-negative function such that for</u> $n = 0,1,2,\ldots$, $x^n W(x) \in L_\infty(\mathbb{R})$. <u>Assume further, that there exist</u> $0 < p_0 < \infty$ <u>and an increasing sequence</u> $\{c_n\}_1^\infty$ <u>of positive numbers such that for</u> $n = 1,2,3,\ldots$, $p_0 \le p \le \infty$ <u>and</u> $P \in \mathcal{P}_n$,

$$(15.20)\quad \|PW\|_{L_p(\mathbb{R})} \le 2\|PW\|_{L_p(-c_n,c_n)}.$$

Let $n \geq 1$ and

(15.21) $\quad P(x) := Ax^n + \dots \in P_n,\ A \neq 0,$

and let

(15.22) $\quad r_n := AE_{n\infty}(W).$

Then for $z \in \mathbb{C}\backslash\{x_{jn}\}_1^n$,

(15.23) $\quad |P(z)/p_{n\infty}(W,z) - r_n| \leq 2\ \{\ \|PW\|^2_{L_\infty(\mathbb{R})} - r_n^2\ \}^{1/2}/d_n(z),$

where $\{x_{jn}\}_1^n$ are the zeros of $T_{n\infty}W,x)$, and $d_n(z)$ is given by (15.5).

Proof

A glance at the proof of Theorem 15.1 shows that, because the sequence $\{c_n\}_1^\infty$ is uniform in p, so is the estimate (15.6). Letting $p \to \infty$, we then obtain (15.23), noting that as $p \to \infty$,

$$\|PW\|_{L_p(\mathbb{R})} \to \|PW\|_{L_\infty(\mathbb{R})};$$

$$E_{np}(W) \to E_{n\infty}(W);$$

$$T_{np}(W,z) \to T_{n\infty}(W,z),\ \text{uniformly in compact subsets of } \mathbb{C}.\ \square$$

Proof of (3.7) and (3.8) when $p = \infty$.

Let $c_n := a_n(1 + K\{\log n/n\}^{2/3})$, $n = 1,2,3,\dots$, where K is a fixed but sufficiently large positive number. By Lemma 7.4, (15.20) is satisfied uniformly for $1 \leq p \leq \infty$. Hence the hypotheses of Lemma 15.2 are satisfied. Now let $P_{n-1}(x)$ be a possibly complex-valued polynomial of degree at most $n - 1$, and let $\hat{S}_{2n-2}(x) := |P_{n-1}(x)|^2$, a polynomial of degree at most $2n - 2$, non-negative in $\mathbb{R}$. By adding, if necessary, a small positive number to $\hat{S}_{2n-2}(x)$, we may assume that it is positive in $\mathbb{R}$. Let

$$S_{2n}(x) := (1 - x^2)\ \hat{S}_{2n-2}(x),$$

and

$$V(x) := \{(1 - x^2)/S_{2n}(x)\}^{1/2} = |P_{n-1}(x)|^{-1}.$$

Further, let $T_n(x)$ be given by (13.8), and

$$P(x) := E_{n\infty}(V)^{-1}\ T_n(x/a_n),$$

which has leading coefficient

$$A := a_n^{-n}\, E_{n\infty}(V)^{-1}.$$

We have by (13.5) and (3.4) for $p = \infty$,

$$(15.24)\qquad r_n := A\, E_{n\infty}(W) = G[V]^{-1}\, G[W(a_n x)]\ (1 + o(1)),\ n \to \infty.$$

Further,

$$\|PW\|_{L_\infty(\mathbb{R})} = \|PW\|_{L_\infty[-a_n,a_n]}$$
$$= E_{n\infty}(V)^{-1}\|T_n(x)W(a_n x)\|_{L_\infty[-1,1]} \le \|P_{n-1}(x)W(a_n x)\|_{L_\infty[-1,1]},$$

by (13.11),(13.5) and the definition of V. Substituting into (15.23), we obtain after setting $z = a_n u$ and noting that all zeros of $T_{n\infty}(W,x)$ lie in $[-a_n,a_n]$,

$$(15.25)\qquad |\ \{T_n(u)/T_{n\infty}(W,a_n u)\}\{E_{n\infty}(W)/E_{n\infty}(V)\} - r_n|$$
$$\le 2\ \{\ \|P_{n-1}(x)W(a_n x)\|^2_{L_\infty[-1,1]} - r_n^2\ \}^{1/2}\ /d(a_n u/c_n,[-1,1]),$$

where d(,) denotes distance between points and/or sets. We now choose $P_{n-1}(x)$ much as in the proof of (3.4) for $p = \infty$. Let h(t) be a polynomial positive in [-1,1], and

$$g(t) := \begin{cases}(1 - t^2)\ h(t), & t \in [-1,1],\\ 0 & ,\ t \in \mathbb{R}\backslash[-1,1].\end{cases}$$

By Theorem 12.3, we can choose polynomials $P_{n-1}(x)$ of degree at most $n - \log n$, such that (14.19) and (14.20) hold. Then as in the proof of (3.4) for $p = \infty$, and by (15.24),

$$r_n = G[g]\ (1 + o(1)),\ n \to \infty.$$

Further,

$$\lim_{n\to\infty} \|P_{n-1}(x)W(a_n x)\|_{L_\infty[-1,1]} = \|g\|_{L_\infty[-1,1]}.$$

Then substituting into (15.25), and using the asymptotics for $E_{n\infty}(W)$ and $E_{n\infty}(V)$, we obtain

$$|\{T_n(u)/T_{n\infty}(W,a_n u)\}\{a_n^n\ G[g]\ (1 + o(1))\} - G[g](1 + o(1))|$$
$$\le 2\ \{\ \|g\|^2_{L_\infty[-1,1]} + o(1) - G[g]^2\ \}\ /\ \epsilon\ ,$$

uniformly for $d(u,[-1,1]) \ge 2\epsilon$, and n large enough. Choosing g approximately 1 in the sense that

$$\|g\|_{L_\infty[-1,1]} \leq 1,$$

and for small positive δ,

$$1 + \delta \geq G[g] \geq 1 - \delta,$$

we obtain uniformly for $d(u,[-1,1]) \geq 2\epsilon$,

$$\limsup_{n \to \infty} \left| T_n(u) / \{ a_n^{-n} T_{n\infty}(W, a_n u) \} - 1 \right| \leq \eta,$$

where η may be made arbitrarily small. The proof may now be completed as it was for the case $1 < p < \infty$. □

16. The Case $p = 2$: Orthonormal Polynomials.

In this section, we use the relationship between orthonormal polynomials on $[-1,1]$ and those on the circle to improve our results in the case $p = 2$. Let $w(x)$ be a non-negative integrable function on $[-1,1]$ and let $p_n(w,x) = \gamma_n(w)x^n + \ldots$, $n = 0,1,2,\ldots$, denote the corresponding orthonormal polynomials. Define the associated measure

$$(16.1)\qquad d\mu(\theta) := w(\cos\theta)\ |\sin\theta|\ d\theta,\ \theta \in [-\pi,\pi],$$

and the associated orthonormal polynomials

$$\phi_n(z) := \phi_n(d\mu,z) := \kappa_n(d\mu)\ z^n + \ldots,\ \kappa_n(d\mu) > 0,\ n = 0,1,2,\ldots,$$

satisfying for $m,n = 0,1,2,\ldots$,

$$(16.2)\qquad (2\pi)^{-1}\int_{-\pi}^{\pi} \phi_n(d\mu,z)\ \overline{\phi_m(d\mu,z)}\ d\mu(\theta) = \delta_{mn},\ z := e^{i\theta}.$$

Further, we define the monic orthogonal polynomials

$$(16.3)\qquad \Phi_n(z) := \Phi_n(d\mu,z) := \phi_n(d\mu,z)/\kappa_n(d\mu),\ n = 0,1,2,\ldots .$$

Szegö [66,p.294] gives the identity

$$(16.4)\qquad p_n(w,x) = (2\pi)^{-1/2}(1 + \Phi_{2n}(0))^{-1/2}\ \{z^{-n}\phi_{2n}(z) + z^n\phi_{2n}(z^{-1})\},$$

which yields via straightforward manipulations

$$(16.5)\qquad \gamma_n(w) = (2\pi)^{-1/2}\ 2^n\ \kappa_{2n}(d\mu)\ (1 + \Phi_{2n}(0))^{1/2}.$$

This last relation enables us to prove

<u>Lemma 16.1</u>

<u>Let</u> $\hat{W}(x)$ <u>be a non-negative function such that</u> $x^n\hat{W}^2(x) \in L_2(\mathbb{R})$, $n = 0,1,2,\ldots$. <u>Let</u> $a > 0$ <u>and</u>

$$(16.6)\qquad d\mu(\theta) := a\ \hat{W}^2(a\cos\theta)\ |\sin\theta|\ d\theta,\ \theta \in [-\pi,\pi].$$

<u>Then for</u> $n = 0,1,2,\ldots$,

$$(16.7)\qquad E_{n2}(\hat{W}) \geq 2^{-n}\ \pi^{1/2}\ (1 + \Phi_{2n}(d\mu,0))^{-1/2}\ a^{n+1/2}\ G[\hat{W}(ax)].$$

Proof

We have

$$E_{n2}(\hat{W}) \geq a^n \min_{P \in \mathcal{P}_{n-1}} \|(u^n - P(u))a^{1/2}\hat{W}(a_n u)\|_{L_2[-1,1]} = a^n \gamma_n(w)^{-1},$$

where

$$w(x) := \begin{cases} a\ \hat{W}^2(ax), & x \in [-1,1], \\ 0 & , x \in \mathbb{R}\backslash[-1,1]. \end{cases}$$

Thus, by (16.5), and with $d\mu$ as in (16.6),

$$E_{n2}(\hat{W}) \geq a^n (2\pi)^{1/2} 2^{-n} \kappa_{2n}(d\mu)^{-1} (1 + \Phi_{2n}(d\mu,0))^{-1/2}.$$

As in Szegö [66,p.301], an application of the arithmetic-geometric mean inequality yields

$$\kappa_{2n}(d\mu)^{-1} = \min_{P \in \mathcal{P}_{2n-1}} \left\{ \frac{1}{2\pi} \int_{-\pi}^{\pi} |z^n - P(z)|^2 \, d\mu(\theta) \right\}^{1/2}$$

$$\geq \exp \left(\frac{1}{4\pi} \int_{-\pi}^{\pi} \log \{w(\cos\theta)\ |\sin\theta|\} \, d\theta\right)$$

$$= a^{1/2} G[w]^{1/2} G[1 - x^2]^{1/4} = a^{1/2} G[\hat{W}(ax)]/\sqrt{2},$$

by (13.23). Hence (16.7). □

We see that the lower bound in (16.7) will be useful, if we can show that $\Phi_{2n}(d\mu,0)$ is negligible. The estimation of this quantity plays a crucial role in recent extensions of Szegö's theory [39, 40, 56, 61] and also in the resolution of Freud's Conjecture. Here we shall use Theorem 2.3 of Knopfmacher, Lubinsky and Nevai [21]:

Lemma 16.2

Let $\hat{W}(x)$ be a non-negative function on $\mathbb{R}$ such that $x^n\hat{W}^2(x) \in L_2(\mathbb{R})$, $n = 0,1,2,\ldots$. Suppose that $\{c_n\}_1^\infty$ is an increasing sequence of positive numbers, and that there exist real polynomials $S_{n-2}(x)$ of degree at most $n - 2$, $n = 2,3,4,\ldots$, such that for $p = 1/2$ and $p = 2$,

$$(16.8) \qquad \lim_{n\to\infty} \pi^{-1} \int_{-1}^{1} |W(c_n x)S_{n-2}(x)(1 - x^2)^{1/4}|^p \, (1 - x^2)^{-1/2} \, dx = 1.$$

Then, if for $n = 1,2,3,\ldots$,

(16.9) $$d\mu_n(\theta) := c_n W^2(c_n \cos\theta)\,|\sin\theta|\,d\theta,\ \theta \in [-\pi,\pi],$$

we have

(16.10) $$\lim_{n\to\infty}\{\sup_{m\geq 2n-3}|\phi_m(d\mu_n,0)|\} = 0.$$

Proof

Theorem 2.3 in [21] states (16.10), under the hypotheses of Theorem 1 in [21]. In addition to (16.8), the hypotheses of Theorem 1 in [21] involve an infinite-finite range inequality. However, a glance at the proof of Theorem 2.3. in [21] shows that the latter was not used in proving (16.10). It was used only later, in obtaining the asymptotics for the recurrence relation coefficients. □

In the proof of Theorem 3.5, we shall need the following crude infinite-finite range inequality:

Lemma 16.3

Let $W(x)$ be as in Theorem 3.1, and let $\hat{W} := W\psi$, where ψ is a non-negative function such that $x^n\hat{W}^2(x) \in L_2(\mathbb{R})$, $n = 0,1,2,\ldots$. Assume further, that

(16.11) $$\lim_{|x|\to\infty} |\log\psi(x)|\,Q(x)^{-1/4} = 0.$$

Then there exists a sequence of positive numbers $\{\epsilon_n\}_1^\infty$ such that

(16.12) $$\lim_{n\to\infty} \epsilon_n n^{1/2} = 0,$$

and such that, if

(16.13) $$c_n := a_n(1+\epsilon_n),\ n = 1,2,3,\ldots ,$$

then for $n \geq 1$ and $P \in \mathcal{P}_n$,

(16.14) $$\|P\hat{W}\|_{L_2(|x|\geq c_n)} \leq 2\,e^{-n^{1/4}/\log n}\,\|PW\|_{L_2(|x|\leq c_n)},$$

and

(16.15) $\quad \|P\hat{W}\|_{L_2(\mathbb{R})} \le \|P\hat{W}\|_{L_2(|x|\le c_n)} + 2e^{-n^{1/4}/\log n}\,\|PW\|_{L_2(|x|\le c_n)}.$

Proof

Let k be a fixed positive integer. For n large enough,

$$Q(a_{kn}) \sim a_{kn}Q'(a_{kn}) \sim n,$$

by (6.20) and (6.21) in Lemma 6.2. Then, from (16.11),

$$\max_{a_n\le|x|\le a_{kn}} |\log \psi(x)| \le \delta_n n^{1/4},$$

where

$$\lim_{n\to\infty} \delta_n = 0,$$

and we may assume that

(16.16) $\quad \delta_n \ge 1/\log n, \; n = 2,3,4,\ldots .$

Now let $\epsilon_n := (K_n \log n / n)^{2/3}$, $n = 2,3,4,\ldots$, where K_n is chosen so that

$$C_2K_n \log n = 2\,\delta_n\, n^{1/4},\; n = 2,3,4,\ldots ,$$

and C_2 is the constant in Theorem 7.2. Further, let $\{c_n\}_1^\infty$ be given by (16.13). Then by Theorem 7.2, for $P \in \mathcal{P}_n$, $n \ge 2$,

$$\|P\hat{W}\|_{L_2(c_n\le|x|\le a_{kn})} \le \exp(\delta_n n^{1/4})\,\|PW\|_{L_2(|x|\ge c_n)}$$

$$\le \exp(\delta_n n^{1/4} - C_2K_n\log n)\,\|PW\|_{L_2(|x|\le c_n)}$$

$$\le \exp(-n^{1/4}/\log n)\,\|PW\|_{L_2(|x|\le c_n)},$$

by choice of K_n and (16.16). It remains to deal with $\|PW\|_{L_2(|x|\ge a_{kn})}$, k some fixed, but large enough, positive integer. There are several standard ways of showing that this is negligible [25,46,62] but nevertheless, we provide some details: Note first, from Lemma 8.5, that a_n is of at most polynomial growth at ∞, that is for some $C > 0$, $a_n \le n^C$, $n \ge 2$. Furthermore, (16.11) shows that if k is large enough,

$$\hat{W}(x) \le W^{1/2}(x),\; |x| \ge a_{kn},\; n \ge 1.$$

Then for $P \in \mathcal{P}_n$, $n \ge 2$, if k is large enough,

$$\|P\hat{W}\|_{L_2(|x|\geq a_{kn})} \leq \|PW^{1/2}\|_{L_2(|x|\geq a_{kn})} = \|P^2W\|^{1/2}_{L_1(|x|\geq a_{kn})}$$

$$\leq a_{kn}^{-n} \|P^2(x)W(x)x^{2n}\|^{1/2}_{L_1(|x|\geq a_{kn})} \leq a_{kn}^{-n} \|P^2(x)W(x)x^{2n}\|^{1/2}_{L_1(|x|\geq ra_{4n})},$$

some fixed $r > 1$, independent of $k \geq 8$, by (8.13) in Lemma 8.5. Then Theorem 7.2 yields

$$\|P\hat{W}\|_{L_2(|x|\geq a_{kn})} \leq a_{kn}^{-n} e^{-C_5 n/\log n} \|P^2(x)W(x)x^{2n}\|^{1/2}_{L_1(|x|\leq ra_{4n})}$$

$$\leq (ra_{4n}/a_{kn})^n e^{-C_5 n/\log n} \|P^2(x)W^2(x)\|^{1/2}_{L_1(|x|\leq ra_{4n})} W^{-1/2}(ra_{4n})$$

$$\leq (C_6 a_{4n}/a_{kn})^n \|PW\|_{L_2(|x|\leq c_n)},$$

by Theorem 7.2, and since $Q(ra_{4n}) \sim n$. Here, C_6 is independent of k. Choosing k large enough, and noting that (8.13) in Lemma 8.5 implies

$$a_{3n/2}/a_n \geq s > 1, \ n \text{ large enough},$$

we see that for $n \geq 2$, $P \in \mathcal{P}_n$,

$$\|P\hat{W}\|_{L_2(|x|\geq a_{kn})} \leq 2^{-n} \|PW\|_{L_2(|x|\leq c_n)}. \quad \square$$

Proof of Theorem 3.5

We first prove the asymptotic upper bound. Throughout this proof, we fix $\{c_n\}_1^\infty$ to be the sequence in Lemma 16.3. By (16.15) and a substitution, we see that

$$(16.17) \quad E_{n2}(\hat{W}) \leq c_n^{n+1/2} \min_{P\in\mathcal{P}_{n-1}} \{\|(u^n - P(u))\hat{W}(c_n u)\|_{L_2[-1,1]}$$
$$+ 2\, e^{-n^{1/4}/\log n} \|(u^n - P(u))W(c_n u)\|_{L_2[-1,1]}\}.$$

Now let $V(x)$ and $V_2(x)$ be as in Theorem 13.1, and let $T_n(x)$ be given by (13.8). Let F be some measurable subset of $[-1,1]$. Substituting $T_n(x)$ in (16.17), and using (13.5), (13.10), and (13.11), we obtain:

$$(16.18) \quad E_{n2}(\hat{W}) \leq c_n^{n+1/2} E_{n2}(V_2) \{ \|V^{-1}(u)\hat{W}(c_n u)\|_{L_2(F)} \sigma_2^{-1}$$
$$+ \|V_2^{-1}(u)\hat{W}(c_n u)\|_{L_\infty([-1,1]\setminus F)}$$

$$+ 2e^{-n^{1/4}/\log n} \, \|V^{-1}(u)W(c_n u)\|_{L_2[-1,1]}/\sigma_2 \}.$$

Now let $g(x)$ be a function positive and continuous in $[-1,1]$. Let $S_n(x)$ be the polynomial of degree k_n, say, $n = 1,2,3,\ldots$, satisfying (3.28), (3.29) and (3.30). By assumption,

$$\lim_{n\to\infty} k_n n^{-1/2} = 0.$$

By Theorem 12.1, we can choose polynomials $\hat{P}_n(x)$ of degree at most $n - k_n - \log n$, satisfying

$$\lim_{n\to\infty} \int_{-1}^{1} |\log |\hat{P}_n(x)W(c_n x)/g(x)|| \, (1 - x^2)^{-1/2} \, dx = 0,$$

and

$$\|\hat{P}_n(x)W(c_n x)\|_{L_\infty[-1,1]} \le C_5, \quad n = 1,2,3,\ldots,$$

and

$$\lim_{n\to\infty} \hat{P}_n(x)W(c_n x) = g(x),$$

uniformly in compact subsets of $\{x: 0 < |x| < 1\}$. Let

$$P_n(x) := \hat{P}_n(x)S_n(x),$$

a polynomial of degree at most $n - \log n$, $n = 1,2,3,\ldots$, satisfying

$$\lim_{n\to\infty} \int_{-1}^{1} |\log |P_n(x)\hat{W}(c_n x)/g(x)|| \, (1 - x^2)^{-1/2} \, dx = 0, \tag{16.19}$$

$$\|P_n(x)W(c_n x)\|_{L_2[-1,1]} \le C_5 \, \|S_n\|_{L_2[-1,1]} = o(e^{n^{1/4}/\log n}), \tag{16.20}$$

and in view of (3.30),

$$\|P_n(x)\hat{W}(c_n x)\|_{L_2[-1,1]} \le C_6, \quad n = 1,2,3,\ldots, \tag{16.21}$$

and

$$\sup_n \|P_n(x)\hat{W}(c_n x)\|_{L_2(E)} \to 0, \text{ as } \operatorname{meas}(E) \to 0. \tag{16.22}$$

Now (16.19) implies that $\{ (P_n(x)\hat{W}(c_n x))^{\pm 1} \}_1^\infty$ converges in measure to $g^{\pm 1}$ in $[-1,1]$. Then, given $\eta, \epsilon > 0$, we can find for each n a set F_n of measure less than ϵ such that

$$\sup_n \|P_n(x)\hat{W}(c_n x)\|_{L_2(F_n)} < \eta, \tag{16.23}$$

and for $n \ge n_0(\epsilon, \eta)$,

(16.24) $\quad 1 - \eta \le |P_n(x)\hat{W}(c_nx)|/g(x) \le 1 + \eta, \ x \in (-1,1)\backslash F_n.$

Next, let S(x) in Theorem 13.1 be $(1 - x^2)|P_n(x)|^2$, a polynomial of degree at most 2(n - log n + 1). By adding a small positive constant, we may ensure that S(x) is positive in (-1,1). Then,

$$V(x) = |P_n(x)|^{-1},$$

and

$$V_2(x) = (1 - x^2)^{-1/4}|P_n(x)|^{-1}.$$

Substituting into (16.18), we obtain

$$\begin{aligned} E_{n2}(\hat{W}) \le c_n^{n+1/2}\ E_{n2}(V_2)\ \{\ &\|P_n(x)\hat{W}(c_nx)\|_{L_2(F_n)}\ \sigma_2^{-1} \\ &+ \|P_n(x)\hat{W}(c_nx)(1 - x^2)^{1/4}\|_{L_\infty((-1,1)\backslash F_n)} \\ &+ e^{-n^{1/4}/\log n}\ \|P_n(x)W(c_nx)\|_{L_2[-1,1]}\sigma_2^{-1}\ \} \\ \le c_n^{n+1/2}\ G[\hat{W}(c_nx)/g(x)]\ 2^{-n+1/2}\ \pi^{1/2}\{\ &\eta\ \sigma_2^{-1} \\ &+ (1 + \eta)\|g(x)(1 - x^2)^{1/4}\|_{L_\infty[-1,1]} + o(1)\ \}, \end{aligned}$$

by (16.19) to (16.24) and by (13.5). Since $\eta > 0$ is arbitrary, and we may choose g(x) to approximate $(1 - x^2)^{-1/4}$ in the sense that

$$G[1/g(x)] \le (1 + \eta)\ G[(1 - x^2)^{1/4}] = (1 + \eta)/\sqrt{2}$$

and

$$\|g(x)(1 - x^2)^{1/4}\|_{L_\infty[-1,1]} \le 1,$$

we obtain

$$\limsup_{n \to \infty} E_{n2}(\hat{W})/\{\ (c_n/2)^{n+1/2}\ G[\hat{W}(c_nx)]\ (2\pi)^{1/2}\ \} \le 1.$$

Now by Lemma 13.4, (16.12) and (16.13), we have

$$c_n^{n+1/2}G[W(c_nx)] = a_n^{n+1/2}G[W(a_nx)]\ (1 + o(1)).$$

Thus if we can show

(16.25) $\quad G[\psi(c_nx)]/G[\psi(a_nx)] \to 1, \ n \to \infty,$

then

(16.26) $\quad \limsup_{n \to \infty} E_{n2}(\hat{W})/\{\ (a_n/2)^{n+1/2}\ G[\hat{W}(a_nx)]\ \} \le (2\pi)^{1/2}.$

Now, as in the proof of Lemma 16.3, we see that

$\log \psi(a_n u) = o(n^{1/4})$, uniformly for $K/a_n \le |u| \le 1$,

and hence for each fixed, but large enough r,

$$(16.27) \qquad \int_{1\ge|u|\ge 1-rn^{-1/2}} \frac{|\log \psi(a_n u)| + |\log \psi(c_n u)|}{(1-u^2)^{1/2}} du$$

$$= o(n^{1/4}) \int_{1\ge|u|\ge 1-rn^{-1/2}} (1-u^2)^{-1/2} du \to 0,\ n \to \infty.$$

Further,

$$(16.28) \qquad \int_{|u|\le r/a_n} \frac{|\log \psi(a_n u)| + |\log \psi(c_n u)|}{(1-u^2)^{1/2}} du = O(1/a_n) \to 0,\ n \to \infty.$$

Finally,

$$I := \int_{r/a_n\le|u|\le 1-rn^{-1/2}} \frac{\log \psi(a_n u) - \log \psi(c_n u)}{(1-u^2)^{1/2}} du$$

$$= \int_{s/a_n\le|x|\le 1-sn^{-1/2}} \log \psi(a_n x) \left\{\frac{1}{(1-x^2)^{1/2}} - \frac{a_n/c_n}{(1-(a_n x/c_n)^2)^{1/2}}\right\} dx$$

$$+ o(1),$$

by (16.27) and (16.28) and the substitution $c_n u = a_n x$. Here s is some fixed positive number. In this last integrand, $\log \psi(a_n x) = o(n^{1/4})$. Further, a straightforward computation shows that

$$\frac{1}{(1-x^2)^{1/2}} - \frac{a_n/c_n}{(1-(a_n x/c_n)^2)^{1/2}} = O(|a_n/c_n - 1|(1-|x|)^{-3/2})$$

$$= o(n^{-1/2}(1-|x|)^{-3/2}).$$

Then

$$I = o(n^{1/4}.n^{-1/2}) \int_{s/a_n\le|x|\le 1-sn^{-1/2}} (1-|x|)^{-3/2} dx + o(1) \to 0,$$

$n \to \infty$. Thus

$$\lim_{n\to\infty} \int_{-1}^{1} \frac{\log \psi(a_n x) - \log \psi(c_n x)}{(1-x^2)^{1/2}} dx = 0,$$

which is equivalent to (16.25). Hence (16.26) holds.

Proof of the lower bound corresponding to (16.26).

Let $F, g, \{P_n\}$ be as at (16.19) to (16.24). We show that $S_{n-2} = P_n$ satisfies a relation similar to (16.8). Now (16.24) shows that for $p = 1/2, 2$,

$$\int_{[-1,1]\setminus F_n} |P_n(x)\hat{W}(c_nx)(1 - x^2)^{1/4}|^p (1 - x^2)^{-1/2} \, dx$$

$$= \int_{[-1,1]\setminus F_n} |g(x)(1 - x^2)^{1/4}|^p (1 - x^2)^{-1/2} \, dx + O(\eta).$$

Further, for $p = 2$, we obtain directly from (16.23),

$$\int_{F_n} |P_n(x)\hat{W}(c_nx)(1 - x^2)^{1/4}|^p (1 - x^2)^{-1/2} \, dx = O(\eta^2),$$

while if $p = 1/2$, an application of Holder's inequality with parameters 4,4/3 shows that this last integral is $O(\eta^{1/2})$. Choosing $g(x)$ to approximate $(1 - x^2)^{-1/4}$ in a suitable sense, we see that we can satisfy (16.8) with $p = 1/2$ and $p = 2$. Then if $d\mu_n$ is given by (16.9), we have (16.10), so Lemma 16.1 yields

$$E_{n2}(\hat{W}) \geq 2^{-n} \pi^{1/2} (1 + o(1)) \, c_n^{n+1/2} \, G[\hat{W}(c_nx)]$$

$$\geq (a_n/2)^{n+1/2} (2\pi)^{1/2} G[\hat{W}(a_nx)] (1 + o(1)),$$

as before. Hence

$$\lim_{n\to\infty} E_{n2}(\hat{W})/\{ (a_n/2)^{n+1/2} (2\pi)^{1/2} G[\hat{W}(a_nx)] \} = 1. \quad \square$$

The proof of the asymptotic behaviour of $T_{n2}(W,z)$ and hence of $p_{n2}(W,z)$ is very similar to that in Theorem 3.1, and so is omitted.

Proof of Theorem 3.4

We must show that $\psi(x) := h(x)w_F(x)$ satisfies the hypotheses of Theorem 3.5. Firstly (3.26) is satisfied since

$$\log \psi(x) \sim \log |x|, \; |x| \text{ large enough},$$

while $Q(x)$ grows at least as fast as some positive power of $|x|$ - see (6.18) and (6.20). Next, we construct $\{S_n\}_1^\infty$ with the properties (3.28) to (3.30).

Let l be a positive integer such that $l > \max_j |\Delta_j|$, and let

$$R(x) := \prod_{j=1}^{N} (x - z_j)^l.$$

Then $R(x)w_F(x)$ is continuous in $\mathbb{R}$ and if

$$L := Nl + \sum_{j=1}^{N} \Delta = N\ell + \Delta,$$

then

$$\lim_{|x|\to\infty} R(x)w_F(x)\ |x|^{-L} = 1.$$

Now let $g(x)$ be a continuous function, positive in $[-2,2]$, and let $Z_n(x)$ be a polynomial of degree at most $\log n$, n large enough, such that

$$\lim_{n\to\infty} \|g(x) - Z_n(x)\|_{L_\infty[-2,2]} = 0.$$

Further, let $S_n(x) := R(c_n)Z_n(x)c_n^{-L}$, n large enough, a polynomial of degree at most $2 \log n$, satisfying

$$\lim_{n\to\infty} \int_{-1}^{1} |\log \{S_n(x)\psi(c_n x)\}|\ (1 - x^2)^{-1/2}\, dx$$

$$= \lim_{n\to\infty} \int_{-1}^{1} |\log \{R(c_n x)w_F(c_n x)c_n^{-L}\ Z_n(x)h(c_n x)\}|\ (1 - x^2)^{-1/2}\, dx$$

$$= \int_{-1}^{1} |\log \{|x|^L\ g(x)\}|\ (1 - x^2)^{-1/2}\, dx,$$

and

$$\|S_n\|_{L_\infty[-2,2]} \le C\ \|g\|_{L_\infty[-2,2]} c_n^{-\Delta}, \qquad n \text{ large enough.}$$

Then (3.29) and (3.30) follow easily. Choosing $g(x)$ to approximate $|x|^{-L}$ in a suitable sense, we see that we can satisfy (3.28). Thus (3.31) to (3.33) hold.

Using a standard integral [6,p.241,no.863.41], and elementary manipulations, we see that as $n \to \infty$,

$$G[h(a_n x)w_F(a_n x)] = G[|a_n x|^\Delta](1 + o(1)) = (a_n/2)^\Delta(1 + o(1)), \qquad n \to \infty.$$

Then (3.23) and (3.24) follow from (3.32) and (3.33).

Since for $|z| < 1$, (see [66,p.277])

$$D(|a_n \cos\phi|^\Delta; z) = a_n^{\Delta/2} D(|\sin\phi|; iz)^\Delta = a_n^{\Delta/2}(1 + z^2)^{\Delta/2} 2^{-\Delta/2},$$

we also deduce (3.25) from (3.34). □

References.

1. **N.I. Achiezer**, Theory of Approximation, (transl. by C.J. Hyman), Ungar, New York, 1956.

2. **W.C. Bauldry**, "Orthogonal Polynomials Associated with Exponential Weights", Ph.D. dissertation, Ohio State University, Columbus, 1985.

3. **W.C. Bauldry, A. Máté, P. Nevai**, Asymptotics for the Solutions of Systems of Smooth Recurrence Equations, to appear in Pacific J. Math.

4. **D. Bessis, C. Itzykson and J.B. Zuber**, Quantum Field Theory Techniques in Graphical Enumeration, Adv. in Appl. Math., **1**(1980), 109 - 157.

5. **Z. Ditzian and V. Totik**, Moduli of Smoothness, Springer-Verlag, Berlin, 1987.

6. **H.B. Dwight**, Tables of Integrals and Other Mathematical Data, Macmillan, New York, 1961.

7. **P. Erdös**, On the Distribution of the Roots of Orthogonal Polynomials, *in* "Proceedings, Conference on Constructive Theory of Functions" (G.Alexits, *et al*, Eds.), pp.145-150, Akademiai Kiado, Budapest, 1972.

8. **G. Freud**, Orthogonal Polynomials, Pergamon Press, Budapest, 1966.

9. **G. Freud**, On Polynomial Approximation with Respect to General Weights, *in* "Functional Analysis and its Applications" (H.G.Garnir, *et al*., Eds.), pp.149-179, Lecture Notes in Mathematics, Vol.**399**, Springer-Verlag, Berlin,1973.

10. **G. Freud**, On Estimations of the Greatest Zeros of Orthogonal Polynomials, Acta Math. Acad. Sci. Hungar., **25**(1974), 99-107.

11. **G. Freud**, On the Greatest Zero of an Orthogonal Polynomial.2, Acta Sci. Math.(Szeged), **36**(1974), 49-54.

12. **G. Freud**, On the Coefficients in the Recursion Formulae of Orthogonal Polynomials, Proc. Roy. Irish Acad. Sect.A, **76**(1976), 1-6.

13. **G. Freud**, On the Zeros of Orthogonal Polynomials with Respect to Measures with Noncompact Support, Anal. Numer. Theor. Approx., **6**(1977), 125-131.

14. **G. Freud**, On Markov-Bernstein-type Inequalities and their Applications, J. Approx. Theory, **19**(1977), 22-37.

15. **G. Freud**, On the Greatest Zero of an Orthogonal Polynomial, J. Approx. Theory, **46**(1986), 16-24.

16. **J.B. Garnett**, Bounded Analytic Functions, Academic Press, New York, 1981.

17. **A.A. Goncar and E.A. Rahmanov**, Equilibrium Measure and The Distribution of Zeros of Extremal Polynomials, Math. USSR. Sbornik, **53**(1986), 119-130.

18. **P. Henrici**, Applied and Computational Complex Analysis, Vol.3, John Wiley, New York, 1986.

19. **A. Knopfmacher and D.S. Lubinsky**, Mean Convergence of Lagrange Interpolation for Freud's Weights with Application to Product Integration Rules, JCAM, **17**(1987), 79-105.

20. **A. Knopfmacher and D.S. Lubinsky**, Analogues of Freud's Conjecture for Erdös Type Weights and Related Polynomial Approximation Problems, *in* "Approximation Theory, Tampa" (E.B. Saff, Ed.), Lecture Notes in Math., Vol. **1287**, Springer-Verlag, 1987, 21-69.

21. **A. Knopfmacher, D.S. Lubinsky and P. Nevai**, Freud's Conjecture and Approximation of Reciprocals of Weights by Polynomials, Constr. Approx., **4**(1988), 9-20.

22. **A.L. Levin and D.S. Lubinsky**, Canonical Products and the Weights $\exp(-|x|^{\alpha})$, $\alpha > 1$, with Applications, J. Approx. Theory, **49**(1987), 149-169.

23. **A.L. Levin and D.S. Lubinsky**, Weights on the Real Line that Admit Good Relative Polynomial Approximation, with Applications, J. Approx. Theory, **49**(1987), 170-195.

24. **G.G. Lorentz**, Approximation of Functions, Holt, Rinehart and Winston, New York, 1966.

25. **D.S. Lubinsky**, A Weighted Polynomial Inequality, Proc. Amer. Math. Soc., **92**(1984), 263-267.

26. **D.S. Lubinsky**, Estimates of Freud-Christoffel Functions for Some Weights with the Whole Real Line as Support, J. Approx. Theory, **44**(1985), 343-379.

27. **D.S. Lubinsky**, Gaussian Quadrature, Weights on the Whole Real Line, and Even Entire Functions with Nonnegative Order Derivatives, J. Approx. Theory, **46**(1986), 297-313.

28. **D.S. Lubinsky**, Even Entire Functions Absolutely Monotone in $(0,\infty)$ and Weights on the Whole Real Line, *in* "Orthogonal Polynomials and Their Applications" (C.Brezinski et al., Eds.), Lecture Notes in Mathematics, Vol.**1171**, Springer-Verlag, Berlin, 1985, pp. 221-229.

29. **D.S. Lubinsky**, Strong Asymptotics for Extremal Errors Associated with Erdös-type Weights, in preparation.

30. **D.S. Lubinsky, H.N. Mhaskar and E.B. Saff**, Freud's Conjecture for Exponential Weights, Bull. Amer. Math. Soc., **15**(1986), 217-221.

31. **D.S. Lubinsky, H.N. Mhaskar and E.B. Saff**, A Proof of Freud's Conjecture for Exponential Weights, Constr. Approx., **4**(1988), 65-83.

32. **D.S. Lubinsky and E.B. Saff**, Uniform and Mean Approximation by Certain Weighted Polynomials, with Applications, Constr. Approx., **4**(1988), 21-64.

33. **D.S. Lubinsky and E.B. Saff**, Sufficient Conditions for Asymptotics Associated with Weighted Extremal Problems on $\mathbb{R}$, to appear in Rocky Mountain Journal of Mathematics.

34. **D.S. Lubinsky and E.B. Saff**, Strong Asymptotics for L_p Extremal Polynomials ($1 < p \leq \infty$) Associated with Weights on (-1,1), *in* "Approximation Theory, Tampa" (E.B. Saff, Ed.), Lecture Notes in Math., Vol. **1287**, Springer-Verlag, 1987, 83-104.

35. **L.S. Luo and J. Nuttall**, Asymptotic Behaviour of the Christoffel Function Related to a Certain Unbounded Set, *in* "Approximation Theory, Tampa" (E.B. Saff, Ed.), Lecture Notes in Math., Vol. **1287**, Springer-Verlag, 1987, 105-116.

36. **Al. Magnus**, A Proof of Freud's Conjecture about Orthogonal Polynomials Related to $|x|^{\rho}\exp(-x^{2m})$, *in* "Orthogonal Polynomials and Their Applications" (C.Brezinski, et al., Eds.), Lecture Notes in Mathematics, Springer-Verlag, Berlin, 1985, 362-372.

37. **Al. Magnus**, On Freud's Equations for Exponential Weights, J. Approx. Theory, **46**(1986), 65-99.

38. **Al. Magnus**, Personal Communication.

39. **A. Máté, P. Nevai and V. Totik**, Asymptotics for the Ratio of Leading Coefficients of Orthonormal Polynomials on the Unit Circle, Constr. Approx., **1**(1985), 63-69.

40. **A. Máté, P. Nevai and V. Totik**, Asymptotics for Orthogonal Polynomials defined by a Recurrence Relation, Constr. Approx., **1**(1985), 231-248.

41. **A. Máté, P. Nevai and V. Totik**, Asymptotics for the Greatest Zeros of Orthogonal Polynomials, SIAM J.Math.Anal., **17**(1986), 745-751.

42. **A. Máté, P. Nevai and V. Totik**, Asymptotics for Zeros of Orthogonal Polynomials Associated with Infinite Intervals, J. London Math. Soc., **33**(1986), 303-310.

43. **A. Máté, P. Nevai and T. Zaslavsky**, Asymptotic Expansion of Ratios of Coefficients of Orthogonal Polynomials with Exponential Weights, Trans. Amer. Math. Soc., **287**(1985), 495-505.

44. **H.N. Mhaskar**, Weighted Analogues of Nikolskii-type Inequalities and their Applications, *in* "Conference in Honor of A.Zygmund", Vol.2, 783-801, Wadsworth, Belmont, Calif., 1983.

45. **H.N. Mhaskar**, Discrepancy Theorems, *in* "Approximation Theory, Tampa" (E.B. Saff, Ed.), Lecture Notes in Math., Vol. **1287**, Springer-Verlag, 1987, 117-131.

46. **H.N. Mhaskar and E.B. Saff**, Extremal Problems for Polynomials with Exponential Weights, Trans. Amer. Math. Soc., **285**(1984), 203-234.

47. **H.N. Mhaskar and E.B. Saff**, Weighted Polynomials on Finite and Infinite Intervals: A Unified Approach, Bull. Amer. Math. Soc., **11**(1984), 351-354.

48. **H.N. Mhaskar and E.B. Saff**, Where does the Sup Norm of a Weighted Polynomial Live?, Constr. Approx., **1**(1985), 71-91.

49. **H.N. Mhaskar and E.B. Saff**, A Weierstrass Type Theorem for Certain Weighted Polynomials, *in* "Approximation Theory and Applications" (S.P. Singh, Ed.), Pitman Publishing Ltd. (1985), 115-123.

50. **H.N. Mhaskar and E.B. Saff**, Where does the L_p Norm of a Weighted Polynomial Live?, Trans. Amer. Math. Soc., **303**(1987), 109-124, (see also the Errata to this paper).

51. **N.I. Muskhelishvili**, Singular Integral Equations, (transl. by J.R.M. Radok), Noordhoff, Groningen, 1953.

52. **P. Nevai**, Some Properties of Orthogonal Polynomials Corresponding to the Weight $(1+x^{2k})^{\alpha}\exp(-x^{2k})$ and their Application in Approximation Theory, Soviet Math. Dokl., **14**(1973), 1116-1119.

53. **P. Nevai**, Orthogonal Polynomials on the Real Line Associated with the Weight $|x|^{\alpha}\exp(-|x|^{\beta})$, 1, Acta Math. Acad. Sci. Hungar., **24**(1973), 335-342.

54. **P. Nevai**, "Orthogonal Polynomials", Memoirs. Amer. Math. Soc., No. **213**(1979).

55. **P. Nevai**, Asymptotics for Orthogonal Polynomials Associated with $\exp(-x^4)$, SIAM J. Math Anal., **15**(1984), 1177-1187.

56. **P. Nevai**, Geza Freud, Orthogonal Polynomials and Christoffel Functions. A Case Study, J. Approx. Theory, **48**(1986), 3-167.

57. **P. Nevai and J.S. Dehesa**, On Asymptotic Average Properties of Zeros of Orthogonal Polynomials, SIAM J. Math. Anal., **10**(1979), 1184-1192.

58. **P. Nevai and V. Totik**, Weighted Polynomial Inequalities, Constr. Approx., **2**(1986), 113-127.

59. **P. Nevai and V. Totik**, Sharp Nikolskii-type Estimates for Exponential Weights, manuscript.

60. **E.A. Rahmanov**, On the Asymptotics of the Ratio of Orthogonal Polynomials, Math. USSR.-Sb., **32**(1977), 199-213.

61. **E.A. Rahmanov**, On the Asymptotics of the Ratio of Orthogonal Polynomials II, Math. USSR.-Sb., **46**(1983), 105-117.

62. **E.A. Rahmanov**, On Asymptotic Properties of Polynomials Orthogonal on the Real Axis, Math. USSR.-Sb., **47**(1984), 155-193.

63. **E.B. Saff**, Incomplete and Orthogonal Polynomials, *in* "Approximation Theory, IV" (C.K.Chui et al., Eds.), pp.219-256, Academic Press, New York, 1983.

64. **R.C. Sheen**, "Orthogonal Polynomials Associated with $\exp(-x^6/6)$", Ph.D. dissertation, Ohio State University, Columbus, 1984.

65. **R.C. Sheen**, Asymptotics for Orthogonal Polynomials Associated with $\exp(-x^6/6)$, J. Approx. Theory, (in press).

66. **G. Szegö**, "Orthogonal Polynomials", Amer. Math. Soc. Colloq. Publ., Vol. **23**, Amer. Math. Soc., Providence, R.I., 1939; 4th ed., 1975.

67. **J.L. Ullman**, Orthogonal Polynomials Associated with an Infinite Interval, Michigan Math. J., **27**(1980), 353-363.

68. **J.L. Ullman**, On Orthogonal Polynomials Associated with an Infinite Interval, *in* "Approximation Theory III" (E.W.Cheney, Ed.), pp.889-895, Academic Press, New York, 1980.

69. **W. Van Assche**, Asymptotic Properties of Orthogonal Polynomials from their Recurrence Formula, 1, J. Approx. Theory, **44**(1985), 258-276.

70. **W. Van Assche**, Weighted Zero Distribution for Polynomials Orthogonal on an Infinite Interval, SIAM J. Math. Anal., **16**(1985), 1317-1334.

71. **W. Van Assche and J.S.Geronimo**, Asymptotics for Orthogonal Polynomials with Regularly Varying Recurrence Coefficients, to appear in Rocky Mt. J. Math.

72. **J.L. Walsh**, Interpolation and Approximation by Rational Functions in the Complex Domain, Amer. Math. Soc. Colloq. Publ. ,Vol. **20**, Amer. Math. Soc., Providence, R.I., (5th edition), 1969.

Subject Index

Vol. 1145: G. Winkler, Choquet Order and Simplices. VI, 143 pages. 1985.

Vol. 1146: Séminaire d'Algèbre Paul Dubreil et Marie-Paule Malliavin. Proceedings, 1983–1984. Edité par M.-P. Malliavin. IV, 420 pages. 1985.

Vol. 1147: M. Wschebor, Surfaces Aléatoires. VII, 111 pages. 1985.

Vol. 1148: Mark A. Kon, Probability Distributions in Quantum Statistical Mechanics. V, 121 pages. 1985.

Vol. 1149: Universal Algebra and Lattice Theory. Proceedings, 1984. Edited by S. D. Comer. VI, 282 pages. 1985.

Vol. 1150: B. Kawohl, Rearrangements and Convexity of Level Sets in PDE. V, 136 pages. 1985.

Vol 1151: Ordinary and Partial Differential Equations. Proceedings, 1984. Edited by B.D. Sleeman and R.J. Jarvis. XIV, 357 pages. 1985.

Vol. 1152: H. Widom, Asymptotic Expansions for Pseudodifferential Operators on Bounded Domains. V, 150 pages. 1985.

Vol. 1153: Probability in Banach Spaces V. Proceedings, 1984. Edited by A. Beck, R. Dudley, M. Hahn, J. Kuelbs and M. Marcus. VI, 457 pages. 1985.

Vol. 1154: D.S. Naidu, A.K. Rao, Singular Pertubation Analysis of Discrete Control Systems. IX, 195 pages. 1985.

Vol. 1155: Stability Problems for Stochastic Models. Proceedings, 1984. Edited by V.V. Kalashnikov and V.M. Zolotarev. VI, 447 pages. 1985.

Vol. 1156: Global Differential Geometry and Global Analysis 1984. Proceedings, 1984. Edited by D. Ferus, R.B. Gardner, S. Helgason and U. Simon. V, 339 pages. 1985.

Vol. 1157: H. Levine, Classifying Immersions into $\mathbb{R}^4$ over Stable Maps of 3-Manifolds into $\mathbb{R}^2$. V, 163 pages. 1985.

Vol. 1158: Stochastic Processes – Mathematics and Physics. Proceedings, 1984. Edited by S. Albeverio, Ph. Blanchard and L. Streit. VI, 230 pages. 1986.

Vol. 1159: Schrödinger Operators, Como 1984. Seminar. Edited by S. Graffi. VIII, 272 pages. 1986.

Vol. 1160: J.-C. van der Meer, The Hamiltonian Hopf Bifurcation. VI, 115 pages. 1985.

Vol. 1161: Harmonic Mappings and Minimal Immersions, Montecatini 1984. Seminar. Edited by E. Giusti. VII, 285 pages. 1985.

Vol. 1162: S.J.L. van Eijndhoven, J. de Graaf, Trajectory Spaces, Generalized Functions and Unbounded Operators. IV, 272 pages. 1985.

Vol. 1163: Iteration Theory and its Functional Equations. Proceedings, 1984. Edited by R. Liedl, L. Reich and Gy. Targonski. VIII, 231 pages. 1985.

Vol. 1164: M. Meschiari, J.H. Rawnsley, S. Salamon, Geometry Seminar "Luigi Bianchi" II – 1984. Edited by E. Vesentini. VI, 224 pages. 1985.

Vol. 1165: Seminar on Deformations. Proceedings, 1982/84. Edited by J. Ławrynowicz. IX, 331 pages. 1985.

Vol. 1166: Banach Spaces. Proceedings, 1984. Edited by N. Kalton and E. Saab. VI, 199 pages. 1985.

Vol. 1167: Geometry and Topology. Proceedings, 1983–84. Edited by J. Alexander and J. Harer. VI, 292 pages. 1985.

Vol. 1168: S.S. Agaian, Hadamard Matrices and their Applications. III, 227 pages. 1985.

Vol. 1169: W.A. Light, E.W. Cheney, Approximation Theory in Tensor Product Spaces. VII, 157 pages. 1985.

Vol. 1170: B.S. Thomson, Real Functions. VII, 229 pages. 1985.

Vol. 1171: Polynômes Orthogonaux et Applications. Proceedings, 1984. Edité par C. Brezinski, A. Draux, A.P. Magnus, P. Maroni et A. Ronveaux. XXXVII, 584 pages. 1985.

Vol. 1172: Algebraic Topology, Göttingen 1984. Proceedings. Edited by L. Smith. VI, 209 pages. 1985.

Vol. 1173: H. Delfs, M. Knebusch, Locally Semialgebraic Spaces. XVI, 329 pages. 1985.

Vol. 1174: Categories in Continuum Physics, Buffalo 1982. Seminar. Edited by F.W. Lawvere and S.H. Schanuel. V, 126 pages. 1986.

Vol. 1175: K. Mathiak, Valuations of Skew Fields and Projective Hjelmslev Spaces. VII, 116 pages. 1986.

Vol. 1176: R.R. Bruner, J.P. May, J.E. McClure, M. Steinberger, H_∞ Ring Spectra and their Applications. VII, 388 pages. 1986.

Vol. 1177: Representation Theory I. Finite Dimensional Algebras. Proceedings, 1984. Edited by V. Dlab, P. Gabriel and G. Michler. XV, 340 pages. 1986.

Vol. 1178: Representation Theory II. Groups and Orders. Proceedings, 1984. Edited by V. Dlab, P. Gabriel and G. Michler. XV, 370 pages. 1986.

Vol. 1179: Shi J.-Y. The Kazhdan-Lusztig Cells in Certain Affine Weyl Groups. X, 307 pages. 1986.

Vol. 1180: R. Carmona, H. Kesten, J.B. Walsh, École d'Été de Probabilités de Saint-Flour XIV – 1984. Édité par P.L. Hennequin. X, 438 pages. 1986.

Vol. 1181: Buildings and the Geometry of Diagrams, Como 1984. Seminar. Edited by L. Rosati. VII, 277 pages. 1986.

Vol. 1182: S. Shelah, Around Classification Theory of Models. VII, 279 pages. 1986.

Vol. 1183: Algebra, Algebraic Topology and their Interactions. Proceedings, 1983. Edited by J.-E. Roos. XI, 396 pages. 1986.

Vol. 1184: W. Arendt, A. Grabosch, G. Greiner, U. Groh, H.P. Lotz, U. Moustakas, R. Nagel, F. Neubrander, U. Schlotterbeck, One-parameter Semigroups of Positive Operators. Edited by R. Nagel. X, 460 pages. 1986.

Vol. 1185: Group Theory, Beijing 1984. Proceedings. Edited by Tuan H.F. V, 403 pages. 1986.

Vol. 1186: Lyapunov Exponents. Proceedings, 1984. Edited by L. Arnold and V. Wihstutz. VI, 374 pages. 1986.

Vol. 1187: Y. Diers, Categories of Boolean Sheaves of Simple Algebras. VI, 168 pages. 1986.

Vol. 1188: Fonctions de Plusieurs Variables Complexes V. Séminaire, 1979–85. Edité par François Norguet. VI, 306 pages. 1986.

Vol. 1189: J. Lukeš, J. Malý, L. Zajíček, Fine Topology Methods in Real Analysis and Potential Theory. X, 472 pages. 1986.

Vol. 1190: Optimization and Related Fields. Proceedings, 1984. Edited by R. Conti, E. De Giorgi and F. Giannessi. VIII, 419 pages. 1986.

Vol. 1191: A.R. Its, V.Yu. Novokshenov, The Isomonodromic Deformation Method in the Theory of Painlevé Equations. IV, 313 pages. 1986.

Vol. 1192: Equadiff 6. Proceedings, 1985. Edited by J. Vosmansky and M. Zlámal. XXIII, 404 pages. 1986.

Vol. 1193: Geometrical and Statistical Aspects of Probability in Banach Spaces. Proceedings, 1985. Edited by X. Fernique, B. Heinkel, M.B. Marcus and P.A. Meyer. IV, 128 pages. 1986.

Vol. 1194: Complex Analysis and Algebraic Geometry. Proceedings, 1985. Edited by H. Grauert. VI, 235 pages. 1986.

Vol. 1195: J.M. Barbosa, A.G. Colares, Minimal Surfaces in $\mathbb{R}^3$. X, 124 pages. 1986.

Vol. 1196: E. Casas-Alvero, S. Xambó-Descamps, The Enumerative Theory of Conics after Halphen. IX, 130 pages. 1986.

Vol. 1197: Ring Theory. Proceedings, 1985. Edited by F.M.J. van Oystaeyen. V, 231 pages. 1986.

Vol. 1198: Séminaire d'Analyse, P. Lelong – P. Dolbeault – H. Skoda. Seminar 1983/84. X, 260 pages. 1986.

Vol. 1199: Analytic Theory of Continued Fractions II. Proceedings, 1985. Edited by W.J. Thron. VI, 299 pages. 1986.

Vol. 1200: V.D. Milman, G. Schechtman, Asymptotic Theory of Finite Dimensional Normed Spaces. With an Appendix by M. Gromov. VIII, 156 pages. 1986.

Vol. 1201: Curvature and Topology of Riemannian Manifolds. Proceedings, 1985. Edited by K. Shiohama, T. Sakai and T. Sunada. VII, 336 pages. 1986.

Vol. 1202: A. Dür, Möbius Functions, Incidence Algebras and Power Series Representations. XI, 134 pages. 1986.

Vol. 1203: Stochastic Processes and Their Applications. Proceedings, 1985. Edited by K. Itô and T. Hida. VI, 222 pages. 1986.

Vol. 1204: Séminaire de Probabilités XX, 1984/85. Proceedings. Edité par J. Azéma et M. Yor. V, 639 pages. 1986.

Vol. 1205: B.Z. Moroz, Analytic Arithmetic in Algebraic Number Fields. VII, 177 pages. 1986.

Vol. 1206: Probability and Analysis, Varenna (Como) 1985. Seminar. Edited by G. Letta and M. Pratelli. VIII, 280 pages. 1986.

Vol. 1207: P.H. Bérard, Spectral Geometry: Direct and Inverse Problems. With an Appendix by G. Besson. XIII, 272 pages. 1986.

Vol. 1208: S. Kaijser, J.W. Pelletier, Interpolation Functors and Duality. IV, 167 pages. 1986.

Vol. 1209: Differential Geometry, Peñíscola 1985. Proceedings. Edited by A.M. Naveira, A. Ferrández and F. Mascaró. VIII, 306 pages. 1986.

Vol. 1210: Probability Measures on Groups VIII. Proceedings, 1985. Edited by H. Heyer. X, 386 pages. 1986.

Vol. 1211: M.B. Sevryuk, Reversible Systems. V, 319 pages. 1986.

Vol. 1212: Stochastic Spatial Processes. Proceedings, 1984. Edited by P. Tautu. VIII, 311 pages. 1986.

Vol. 1213: L.G. Lewis, Jr., J.P. May, M. Steinberger, Equivariant Stable Homotopy Theory. IX, 538 pages. 1986.

Vol. 1214: Global Analysis – Studies and Applications II. Edited by Yu.G. Borisovich and Yu.E. Gliklikh. V, 275 pages. 1986.

Vol. 1215: Lectures in Probability and Statistics. Edited by G. del Pino and R. Rebolledo. V, 491 pages. 1986.

Vol. 1216: J. Kogan, Bifurcation of Extremals in Optimal Control. VIII, 106 pages. 1986.

Vol. 1217: Transformation Groups. Proceedings, 1985. Edited by S. Jackowski and K. Pawalowski. X, 396 pages. 1986.

Vol. 1218: Schrödinger Operators, Aarhus 1985. Seminar. Edited by E. Balslev. V, 222 pages. 1986.

Vol. 1219: R. Weissauer, Stabile Modulformen und Eisensteinreihen. III, 147 Seiten. 1986.

Vol. 1220: Séminaire d'Algèbre Paul Dubreil et Marie-Paule Malliavin. Proceedings, 1985. Edité par M.-P. Malliavin. IV, 200 pages. 1986.

Vol. 1221: Probability and Banach Spaces. Proceedings, 1985. Edited by J. Bastero and M. San Miguel. XI, 222 pages. 1986.

Vol. 1222: A. Katok, J.-M. Strelcyn, with the collaboration of F. Ledrappier and F. Przytycki, Invariant Manifolds, Entropy and Billiards; Smooth Maps with Singularities. VIII, 283 pages. 1986.

Vol. 1223: Differential Equations in Banach Spaces. Proceedings, 1985. Edited by A. Favini and E. Obrecht. VIII, 299 pages. 1986.

Vol. 1224: Nonlinear Diffusion Problems, Montecatini Terme 1985. Seminar. Edited by A. Fasano and M. Primicerio. VIII, 188 pages. 1986.

Vol. 1225: Inverse Problems, Montecatini Terme 1986. Seminar. Edited by G. Talenti. VIII, 204 pages. 1986.

Vol. 1226: A. Buium, Differential Function Fields and Moduli of Algebraic Varieties. IX, 146 pages. 1986.

Vol. 1227: H. Helson, The Spectral Theorem. VI, 104 pages. 1986.

Vol. 1228: Multigrid Methods II. Proceedings, 1985. Edited by W. Hackbusch and U. Trottenberg. VI, 336 pages. 1986.

Vol. 1229: O. Bratteli, Derivations, Dissipations and Group Actions on C*-algebras. IV, 277 pages. 1986.

Vol. 1230: Numerical Analysis. Proceedings, 1984. Edited by J.-P. Hennart. X, 234 pages. 1986.

Vol. 1231: E.-U. Gekeler, Drinfeld Modular Curves. XIV, 107 pages. 1986.

Vol. 1232: P.C. Schuur, Asymptotic Analysis of Soliton Problems. VIII, 180 pages. 1986.

Vol. 1233: Stability Problems for Stochastic Models. Proceedings, 1985. Edited by V.V. Kalashnikov, B. Penkov and V.M. Zolotarev. VI, 223 pages. 1986.

Vol. 1234: Combinatoire énumérative. Proceedings, 1985. Edité pa G. Labelle et P. Leroux. XIV, 387 pages. 1986.

Vol. 1235: Séminaire de Théorie du Potentiel, Paris, No. 8. Directeurs M. Brelot, G. Choquet et J. Deny. Rédacteurs: F. Hirsch et G Mokobodzki. III, 209 pages. 1987.

Vol. 1236: Stochastic Partial Differential Equations and Applications Proceedings, 1985. Edited by G. Da Prato and L. Tubaro. V, 257 pages. 1987.

Vol. 1237: Rational Approximation and its Applications in Mathematics and Physics. Proceedings, 1985. Edited by J. Gilewicz, M. Pindor and W. Siemaszko. XII, 350 pages. 1987.

Vol. 1238: M. Holz, K.-P. Podewski and K. Steffens, Injective Choice Functions. VI, 183 pages. 1987.

Vol. 1239: P. Vojta, Diophantine Approximations and Value Distribution Theory. X, 132 pages. 1987.

Vol. 1240: Number Theory, New York 1984–85. Seminar. Edited by D.V. Chudnovsky, G.V. Chudnovsky, H. Cohn and M.B. Nathanson. V, 324 pages. 1987.

Vol. 1241: L. Gårding, Singularities in Linear Wave Propagation. III, 125 pages. 1987.

Vol. 1242: Functional Analysis II, with Contributions by J. Hoffmann-Jørgensen et al. Edited by S. Kurepa, H. Kraljević and D. Butković. VII, 432 pages. 1987.

Vol. 1243: Non Commutative Harmonic Analysis and Lie Groups. Proceedings, 1985. Edited by J. Carmona, P. Delorme and M. Vergne. V, 309 pages. 1987.

Vol. 1244: W. Müller, Manifolds with Cusps of Rank One. XI, 158 pages. 1987.

Vol. 1245: S. Rallis, L-Functions and the Oscillator Representation. XVI, 239 pages. 1987.

Vol. 1246: Hodge Theory. Proceedings, 1985. Edited by E. Cattani, F. Guillén, A. Kaplan and F. Puerta. VII, 175 pages. 1987.

Vol. 1247: Séminaire de Probabilités XXI. Proceedings. Edité par J. Azéma, P.A. Meyer et M. Yor. IV, 579 pages. 1987.

Vol. 1248: Nonlinear Semigroups, Partial Differential Equations and Attractors. Proceedings, 1985. Edited by T.L. Gill and W.W. Zachary. IX, 185 pages. 1987.

Vol. 1249: I. van den Berg, Nonstandard Asymptotic Analysis. IX, 187 pages. 1987.

Vol. 1250: Stochastic Processes – Mathematics and Physics II. Proceedings 1985. Edited by S. Albeverio, Ph. Blanchard and L. Streit. VI, 359 pages. 1987.

Vol. 1251: Differential Geometric Methods in Mathematical Physics. Proceedings, 1985. Edited by P.L. García and A. Pérez-Rendón. VII, 300 pages. 1987.

Vol. 1252: T. Kaise, Représentations de Weil et GL_2 Algèbres de division et GL_n. VII, 203 pages. 1987.

Vol. 1253: J. Fischer, An Approach to the Selberg Trace Formula via the Selberg Zeta-Function. III, 184 pages. 1987.

Vol. 1254: S. Gelbart, I. Piatetski-Shapiro, S. Rallis. Explicit Constructions of Automorphic L-Functions. VI, 152 pages. 1987.

Vol. 1255: Differential Geometry and Differential Equations. Proceedings, 1985. Edited by C. Gu, M. Berger and R.L. Bryant. XII, 243 pages. 1987.

Vol. 1256: Pseudo-Differential Operators. Proceedings, 1986. Edited by H.O. Cordes, B. Gramsch and H. Widom. X, 479 pages. 1987.

Vol. 1257: X. Wang, On the C*-Algebras of Foliations in the Plane. V, 165 pages. 1987.

Vol. 1258: J. Weidmann, Spectral Theory of Ordinary Differential Operators. VI, 303 pages. 1987.